RETAIL IMPACT ASSESSMENT

The rapid advances in retailing during the latter part of the twentieth century have led to pressures for new forms of shopping development and conflicts with established planning policy on the location of new stores and centres. In response to these pressures, various methods have been devised to assess the potential impact of new retail developments on existing centres. This book reviews the methodology and emphasises a recommended best practice approach to the application of retail impact assessment, which meets the requirements of government policy on retailing.

The book shows why and how the current approaches to retail impact assessment need to be improved. It includes examples from all the major retail sectors, for example food stores, retail warehouses, factory outlet centres and regional shopping centres. Evidence is given of the impacts different types of retail development have and an analysis is made of retail impact issues in planning decisions. Case studies show experiences of retail impact issues in Europe and North America, as well as the UK. Advice is given on the application of best practice, with an illustration of the impact assessment framework and a checklist for retail impact assessment. Conclusions are drawn on policy issues in retail planning and on the importance of a practical guide for assessing retail impact.

The approach recommended has been tested through 12 years of practical experience in planning consultancy, specialising in retail planning. It has proved to be a reliable method of predicting the impact of new retail developments and is sufficiently robust to cope with any future changes in government policy. The book has chapter summaries, as well as a glossary and bibliography. *Retail Impact Assessment* is an introduction to the subject for planners and surveyors working in this field and is a practical guide for new and experienced professionals involved with proposals for new retail development. It is also a valuable text for students studying retail planning.

Dr John England is a Partner in England & Lyle, Chartered Town Planners. He previously worked as a planner with W.S. Atkins and in local government. He has specialised in retail planning and was awarded a PhD in 1997 for research on retail impact assessment.

RICS ISSUES IN REAL ESTATE AND HOUSING
Series Editor: Stephen Brown,
Managing Editor, RICS

Including both leading research and student textbooks, the RICS book series advances an understanding of contemporary issues and debates relating to real estate and housing in both national and global markets.

EUROPEAN INTEGRATION AND HOUSING POLICY
*Edited by Mark Kleinman, Walter Matznetter and
Mark Stephens*

HOUSING, INDIVIDUALS AND THE STATE
The morality of government intervention
Peter King

RETAIL IMPACT ASSESSMENT
A guide to best practice
John England

RETAIL IMPACT ASSESSMENT

A guide to best practice

John England

Routledge
Taylor & Francis Group

LONDON AND NEW YORK

First published 2000
by Routledge
2 Park Square, Milton Park, Abingdon, Oxon, OX14 4RN

Simultaneously published in the USA and Canada
by Routledge
270 Madison Ave, New York NY 10016

Reprinted 2004

Transferred to Digital Printing 2006

Routledge is an imprint of the Taylor & Francis Group

© 2000 John England

The right of John England to be identified as the Author of this Work
has been asserted by him in accordance with the Copyright, Designs
and Patents Act 1988

Typeset in Garamond by
The Running Head Limited, Cambridge

British Library Cataloguing in Publication Data
A catalogue record for this book is available
from the British Library

Library of Congress Cataloging in Publication Data
England, J.R.
Retail impact assessment: a guide to best practice / John England.
p. cm. – (Routledge/RICS issues in real estate & housing series)
Includes bibliographical references and index.
1. Store location. 2. Retail trade–Planning. 3. Central business
districts. I. Title. II. Series.
HF5429.275.E54 2000   00-027471
658.8'7—dc21

ISBN 0–415–21666–4

CONTENTS

CONTENTS

FIGURES

TABLES

ABBREVIATIONS

BDP	Building Design Partnership
BID	business improvement district
CBD	central business district
CCG	community conservation guidance
CES	Centre for Environmental Studies
DCPN	Development Control Policy Note
DETR	Department of the Environment, Transport and the Regions
DoE	Department of the Environment
EDC	Economic Development Committee
EIA	environmental impact assessment
NEDO	National Economic Development Office
NPPG	National Planning Policy Guideline
OXIRM	Oxford Institute for Retail Management
PPG	Planning Policy Guidance
PRAG	Planning Research Applications Group
RIA	retail impact assessment
RICS	Royal Institution of Chartered Surveyors
RTPI	Royal Town Planning Institute
TIA	traffic impact assessment
TRICS	trip rate information computer system
UDP	unitary development plan
URBED	Urban and Economic Development Group
URPI	Unit for Retail Planning Information

PREFACE

I have been involved with retail planning and policy for almost 30 years. During this time there have been many significant changes in retailing; government and planners, too, have changed their attitudes to new shopping development. Retail planning is a rapidly changing field and one in which there has been considerable debate about retail trends and the effects of new forms of retailing on existing centres.

Since moving from local government planning into consultancy in the late 1980s, I have carried out a large number of retail impact assessments for clients, both developers and local authorities. I became sceptical of some of the approaches being used and the quality of the conclusions being drawn from retail impact studies, and therefore of the quality of the advice used in decision-making by local authorities and planning inspectors.

I became aware that my own knowledge was lacking in some areas and I decided to embark on a detailed examination of the application of retail impact assessment in urban planning. I was accepted by Newcastle University in 1994 to study part-time for a PhD, and my doctorate was awarded in 1997. In preparing the thesis and this book, which is closely based on it, my consultancy background has given me access to information from other planning consultants and I have been able to draw on personal contacts with numerous consultants who have experience in retail impact assessment. I have also been able to gain from my personal experience of carrying out retail impact studies and advising clients on shopping issues. The material that has been produced from this consultancy work is used extensively in the book.

I am particularly grateful to Barry Wood of the Department of Town and Country Planning at Newcastle University for his helpful and patient supervision of my PhD research, and to other members of the department – Patsy Healey, Ken Willis, Allan Gillard and Angela Hull – for their assistance during the research.

Acknowledgements are due to Ian Lyle, my business partner in England & Lyle, for his support during the preparation of the thesis and the book. The co-operation of many local authority planners and consultants is greatly appreciated in the development of the ideas which have led to the advice on best practice. I am also grateful to Stephen Brown at the RICS for his positive response to my proposal to write the book in the Issues in Real Estate and Housing series.

Special thanks are due to Joyce who encouraged me to develop the ideas in my thesis into a book, and to my late wife, Sue, and my daughters Elizabeth and Rebecca, for their support when I was writing a thesis at the same time as setting up a planning consultancy business. The effort has been very worthwhile. It has enabled me to turn an interest in retail planning into a motivation to see the need for an improvement in retail impact assessment become a reality.

1

INTRODUCTION

The methodology of retail impact assessment (RIA) has evolved over the last 30 years and has moved through several stages of development. At the same time there have been new directions in planning theory which have in turn influenced planning policy. Shifts have taken place in government policy towards retail development along with changes in attitudes towards new forms of retailing, particularly in out-of-centre locations. A key issue in retail planning is whether major shopping developments have an unaccept-able impact on existing town centres. In the 1990s this question achieved high political profile and has become more significant because of growing public concern about the cycle of decline perceived in many town and city centres.

Approaches to assessing retail impact have changed considerably over recent decades because of technical advances in planners' understanding of the retail system and through learning from past experience of the effects of new retail developments. But, at the same time there has been a realisation that assessing the impact of a new shopping development is not simple; it is concerned with outcomes which cannot easily be predicted or quantified. Human behaviour and the retail system are too complex for RIA to be treated as a mechanistic exercise.

Developments which have taken place in RIA methodology have tended to follow changes in the policy context in which planning is carried out in Britain, and the application of RIA needs to respond further to the policy climate underlying urban planning in the 1990s. A best practice guide is needed to advise those who are involved in assessing the impact of new shopping developments to use a sound methodology and make informed judgements as a basis for planning decisions. Advice on best practice would be very beneficial to local authorities, consultants, retailers, developers and planning inspectors. It is clear, however, that such advice must be firmly rooted in policy and must help in implementing and interpreting current planning policy guidance. Policy guidance on retailing has undergone some radical changes during the 1990s. Development plans must provide a clear policy base for new shopping development, and planning applications and appeals must be decided on the basis of good advice on retail impact issues.

This book therefore sets out the fundamental concepts on which retail planning is based. It seeks to show how the current methodology of RIA could be improved, not just technically but also in order to reflect the requirements of current policy. Consideration is also given to ways in which the application of RIA in the future may have to respond to anticipated trends in retail policy and emerging issues in planning. It is intended to be a critical examination of both the methodology and its application. The focus on RIA needs to be set within a wider context of urban planning in general and retail planning in particular.

The first few chapters deal with the following aspects which are crucial to an understanding of retail impact issues and the application of RIA:

- *background* – the nature of retail impact and its importance in urban planning and the evolution of approaches to RIA from the 1960s through to the 1990s
- *policy context* – a review of retailing in the context of planning theory; current issues in retail planning; government policy guidance and the concept of vitality and viability
- *the conventional methodology of RIA* – an overview of the methodological approaches to RIA and an evaluation of the approaches.

The remaining chapters deal with five substantive areas of analysis:

- *development of a framework for RIA* – the need for improvement in the application of RIA; the development of a recommended approach to best practice; an analysis of capacity and quantitative need; a review of data and assumptions; expenditure flows, and the assessment of quantitative impact
- *qualitative factors* – town centre health check appraisals; the question of qualitative need; the sequential approach; accessibility and sustainability issues; environmental impacts, and interpreting the significance of retail impact
- *evidence of retail impact* – lessons to be learnt on the impact of foodstores, retail warehouses and parks, factory outlet centres and regional shopping centres
- *retail impact in the planning process in Britain* – the views of local authorities on RIA, and an analysis of factors in planning appeal decisions on major retail developments
- *experience in Europe and North America* – international comparisons of the pattern of retail development; the planning response; evidence of impact, and approaches to RIA.

The literature on retail planning is wide-ranging and from diverse academic origins – urban geography, town planning, development economics, estate management, marketing and other disciplines. Six phases can be identified in the development of research and knowledge on retail planning and impact assessment:

- The period from the early 1960s up to the mid-1970s was dominated by publications on shopping models, focusing on the development of models and the problems associated with them. This was also a period of related theoretical developments in terms of rational planning. There are several references in these early publications to experience of new forms of retailing in the USA and Europe.
- The mid-1970s to the early 1980s was the era of post hoc impact studies of superstores and hypermarkets. At this time predictive impact assessments developed and there are numerous sources on developments in RIA methodology concerning superstores.
- In the early to mid-1980s there was great concern with retail data problems and with planning theory, reflecting the different perspectives of free market and neo-Marxist ideologies. Refinements took place in RIA methodology and concern began to be shown about the impact of retail warehouses as well as superstores.
- In the late 1980s policy issues became more prominent with the publication of the government's first Planning Policy Guidance Note 6 (PPG6) and some important reports on prospects for shopping in the future. Further information was produced on shopping in the USA and Europe and the first evidence became available on out-of-town regional shopping centres.
- The early 1990s saw a considerable growth in literature on retail planning, including references to planning theory, policy and methodology. Further research was published on North America and on the new regional shopping centres.
- From 1994 onwards the literature tends to concentrate on policy issues arising from PPG6 and on related matters, for instance vitality and viability of town centres and sustainability. There are numerous publications on new types of retailing such as factory outlet centres.

The Bibliography in this book covers the available literature on retail planning which is relevant to RIA. A more comprehensive bibliography of retail planning has been compiled by the Institute for Retail Studies at the University of Stirling and was published by the National Retail Planning Forum in February 1999.

The terminology of retail planning and RIA is quite specific and it is important to be clear about the terms used and their meaning. The main terms are defined in the Glossary at the end of this book.

The first requirement in discussing RIA is to define *retail impact* and the types of impact. The planning system is concerned essentially with the effects of new or proposed retail developments on existing shopping centres. These effects are usually regarded in terms of their economic impact on the level of trade in shopping centres but there can also be social and environmental impacts. Environmental impact is becoming increasingly important because of the issue of sustainability of new shopping developments and their effect on travel patterns.

It is conventional in retail planning to divide the retail market into two sectors — convenience and comparison. *Convenience* shopping takes place in supermarkets, superstores, specialist food shops, newsagents and local convenience shops. The distinction between supermarkets and superstores is essentially one of size. A superstore is defined as having more than 2,500 square metres of trading floorspace. Large superstores of more than 5,000 square metres of trading floorspace are commonly referred to as hypermarkets. The larger floor area of hypermarkets is usually devoted to the sale of comparison goods.

Comparison shopping takes place in town centre stores and shop units and in a variety of modern, purpose-built outlets — retail warehouses, retail parks, factory outlet centres and regional shopping centres. The bulky goods sector generally operates from retail warehouses and retail parks, but town centres also have bulky goods outlets selling furniture, electrical goods, etc. The definition of 'retail warehouses' refers to their use for the sale of household goods (such as carpets, furniture and electrical goods) and bulky DIY items. Retail parks are agglomerations of at least three retail warehouses.

Retail trade can be measured using definitions of expenditure by goods or business type. The distinction may seem academic but it has practical significance for the methodology of RIA, particularly for convenience shopping. Expenditure by convenience goods type has a relatively low growth rate and applies only to those items sold in superstores and supermarkets which are truly convenience goods. Expenditure by convenience business type, however, has a higher growth rate and applies to all items sold in superstores and supermarkets, including comparison goods. Decisions about which definition to use depend very much on the type of shopping provision in the local situation. The merits of the two alternatives are discussed in Chapter 5.

The definition of *centres* is also important in policy terms. PPG6 identifies various types of centre: local centres, district shopping centres, town centres, and regional shopping centres. Of these, *town centres* are the most significant in relation to planning policy since the primary objective of government policy on retailing is to sustain and enhance the vitality and viability of town centres. The term 'town centre' is used generally to cover city, town and suburban district centres which provide a broad range of facilities and services and act as a community focus. It excludes small parades of shops of purely local significance.

4

PPG6 defines types of *location* which are also important in policy terms: edge-of-centre, out-of-centre, and out-of-town. 'Edge-of-centre' for shopping purposes is taken to be a location within easy walking distance (200–300 metres) of the primary shopping area. Several other terms are used extensively throughout this book and they are briefly introduced here to assist in understanding the subject matter of the following chapters:

- vitality and viability
- major or large-scale retail development
- cumulative effects
- linked trips
- sustainable development.

Vitality and viability are normally combined as a single concept which represents the 'health' of a town centre. It is related to the amount of retail activity which takes place (the number of shoppers that are attracted to a centre) and the amount of trade in a centre (its commercial performance). The measurement of vitality and viability is complex and is a matter which is covered in detail in Chapter 3. Figure 1 of PPG6 (shown in this book as Table 3.1) lists key indicators for measuring vitality and viability, and this book makes recommendations on best practice in carrying out town centre appraisals of vitality and viability.

There is no clear definition of *major or large-scale retail development*, but PPG6 states that retail developments of over 2,500 square metres gross floorspace should be supported by evidence of impact. What is major or large-scale will vary according to its location – urban or rural. A proposal for new retail development should be assessed in relation to the size and function of nearby centres.

PPG6 refers to the need to consider the likely *cumulative effects* of recently completed developments and outstanding planning permissions. The Secretary of State for the Environment has directed local planning authorities to notify him of proposals for major retail development before granting planning permission if there are issues of cumulative impact. He also has the power to call in planning applications to make his own decisions. Therefore, in assessing the impact of a proposed development, it may be necessary also to assess the cumulative effects of other recent developments or proposals in the same locality.

The concept of *linked trips* also features strongly in PPG6. The reasoning is that development in or on the edge of town centres can enable one car journey to serve several purposes, thereby reducing the need to travel and to rely upon the car. This is consistent with the aims of sustainable development. In the case of superstores and supermarkets, town centre and edge-of-centre stores can facilitate a higher proportion of linked trips. Hence, PPG6 favours edge-of-centre sites over those which are out-of-centre.

Planning Policy Guidance Note 1 (PPG1), 'general policy and principles', states that sustainable development seeks to deliver the objective of achieving, now and in the future, economic development to secure higher living standards while protecting and enhancing the environment. The most commonly used definition of *sustainable development* is 'ensuring a better quality of life for everyone, now and for generations to come'. The objectives of PPG6 seek to promote sustainable development and are consistent with the government's sustainable development strategy. Development in town centres is regarded as helping to promote sustainable development.

In the context of large retail developments, sustainable development raises several issues concerning transport:

- the use of cars for shopping trips
- the location of retail development in terms of its accessibility by public transport as well as by car
- the effects of new developments on trip mileage.

These issues are discussed in Chapter 6.

The methodology of RIA also uses specific terms, e.g. trade draw, clawback, trade diversion, and residual turnover. These terms are defined in the Glossary.

2

BACKGROUND

What is retail impact?

The relevance of retail impact to urban planning lies in the need to assess the effects of new or proposed retail developments on existing shopping centres. The definition of retail impact and the way it is measured have been refined over recent decades and only by the late 1990s was there a generally accepted view of what retail impact means and how it should be interpreted.

In their literature review for the Department of the Environment (DoE) on the effects of major out-of-town retail development, Building Design Partnership (BDP) Planning and the Oxford Institute for Retail Management (OXIRM) considered the conceptual and definitional issues concerning retail impact. Their report states:

> Impact is of legitimate concern, it can be argued, for five principal reasons:
>
> - Understanding the effects of change: any change in an economy or a physical environment is of legitimate concern to constituent institutions, organisations or individuals
> - Control of public costs: unregulated private actions may give rise to undesirable public or environmental costs (such as an effect upon transport infrastructure)
> - The efficiency argument: the planning system is concerned with the efficient use and allocation of resources, particularly land
> - The equity argument: the degree of accessibility of different types of retail outlet and of shopping centres directly affects the standard of living of all consumers
> - The quality of life argument: the degree of accessibility of different types of retail outlet and of shopping centres indirectly affects the quality of life of individuals and groups in society (through changes in the quality of town centres and other places where people shop).
>
> (BDP/OXIRM, 1992: 33)

7

It is significant to note how these factors relate to the emphasis of current government policy on the vitality and viability of town centres and on sustainability. The factors listed are concerned essentially with safeguarding the well-being of town centres and existing private and public investment. Only the equity and quality of life arguments, regarding accessibility, can be said to have an environmental objective. In the last few years the focus of impact in terms of government policy has shifted much more towards impact in environmental terms.

Types of impact

Noel (1989) states that it is possible to identify three different types of impact studies on the basis of their application:

- post hoc studies of the trading patterns of new shopping developments
- shopping models, for analysis and projection
- predictive impact assessments of proposed retail developments.

Predictive assessments are regarded as the most controversial because they require estimations of turnover and trading impact (Noel, 1989). The major concern in this book is with predictive impact assessments.

In their discussion of the impact of out-of-centre retailing, Bromley and Thomas (1993: 151) summarise the negative and positive impacts. *Negative* impacts are identified as:

- the decline in the competitive position of many existing centres and associated problems of environmental deterioration
- the inability of the less mobile sections of the community to make full use of the new out-of-centre shopping facilities
- problems of traffic congestion.

On the *positive* side, three principal advantages of the new retail facilities are also identified:

- they offer a wider range of shopping opportunities, appealing especially to the car-borne population
- the competitive threat of the new facilities has led to the refurbishment and revitalisation of some traditional centres
- they have deflected some traffic from town centres and contained or even reduced congestion.

Planning Policy Guidance Note 6 (PPG6), the government's policy guidance on town centres and retail developments, states that all applications for major retail developments (those over 2,500 square metres gross floorspace) should be supported by evidence of their likely economic and other impacts on shopping centres (DoE, 1996, para. 4.13).

Three types of impact can be identified – economic, social and environmental.

- *Economic impact* is concerned fundamentally with the consideration of changes in retail turnover or trading patterns in shopping centres as a result of new shopping developments. Economic factors influence the growth of retail expenditure and indicators such as turnover per square metre of retail floorspace. The employment effects of new development can also be relevant.
- *Social impact* is concerned with demographic and behavioural change. It requires a sociological perspective of the social role of town centres and the implications of new forms of retail development on shopper profiles. It also takes account of issues of disadvantaged shoppers.
- *Environmental impact* is concerned with the effects upon the environment as a result of new retail development, particularly in terms of traffic implications. It also raises issues of the sustainability of new shopping developments.

In its study, 'The future of the high street', the Distributive Trades Economic Development Committee (EDC) (1988: 4) stated that 'finding a balance between the social and environmental as against economic considerations on the location of retail investment should continue to be the objective of central and local government planning policy'. The latest version of PPG6 (DoE, 1996) continues to place emphasis on economic factors but increases the priority to be given to transport and other environmental factors.

Economic impact

New forms of retail development and the pressures they impose on the planning system are led by demand from retailers (or developers acting to promote the interests of the retail industry). Retailers constantly attempt to increase their market share by seeking new locations or increasing the competitive position of existing stores.

Retailing is an economic activity and its success or failure must be measured in economic terms such as investment confidence. The Distributive Trades EDC study acknowledged that the amounts, direction and reasons for investment in shops are key factors affecting the viability of high streets, and the structure and success of high street shopping areas depend largely on both recent investment and future expectations.

> Investment in shops in shopping areas is of benefit to the consumer, retail businesses and their employees, by creating the type of capacity

which is in keeping with modern demand. It replaces older, less efficient methods or practices, and enhances shopping area attraction, often making centres where investment has not taken place appear dated and drab.

(Distributive Trades EDC, 1988: 33)

It will be shown later that much of the planning merits of new retail developments are assessed in relation to their effects on the vitality and viability of existing centres. Economic factors, therefore, have become critical in assessing proposals for new development. Economic analysis is the most frequent form of research undertaken by local authorities to underpin retail policy, usually in response to a specific site proposal. Examples of comprehensive research into social and environmental impacts are limited (BDP/OXIRM, 1992).

Another key element of economic impact is *employment*. The trend towards out-of-town retailing has raised concerns about the adverse employment effects of new retail developments. Local planning authorities have often questioned the *net* employment generation of new developments, particularly superstores, though there appears to be little evidence of substantial negative impact arising from the development of superstores. New retail developments result in job creation through direct employment in retail facilities and indirect employment in providing goods and services to the new store. Jobs are also created during construction. But what is the extent of job losses elsewhere if a new development diverts trade and results in closures of other shops? Concern has also been expressed by local planning authorities about the types of jobs created in new retail facilities, many of which are part-time or casual in nature.

Social impact

Social impacts are reflected in changes to the diversity and variety of shopping opportunities in town centres, demonstrated through the threat to the small shop, changes in the importance of non-retail functions in town centres, and increases in social malaise such as crime and vandalism.

(BDP/OXIRM, 1992: 61)

Concern with social impact arises from the view that new development benefits the more mobile, more affluent shopper. Other groups may be disadvantaged. A study by the Royal Town Planning Institute's (RTPI) Retail Planning Working Party examined this issue. Its report states that:

Competition in the market place will normally result in retailers providing for various groups of consumers the goods and services they require, and in locations which are accessible to them. Nevertheless some consumers, because of limited resources or health and associated lack of mobility, are unable to exercise choice and are thus disadvantaged.

(RTPI, 1988: 35)

The same study identified a number of different (but often overlapping) groups whose members are more likely to be disadvantaged as shoppers than the general population. These are:

- low income earners
- residents of locations poorly served by public transport, particularly on peripheral estates or in rural areas
- those without access to a car for routine shopping trips
- those with caring responsibilities (more frequently women), for young children or elderly relatives
- the disabled and others with mobility problems
- the young
- ethnic minorities (RTPI, 1988: 32).

Whilst some groups are disadvantaged, it should also be noted more positively that shopping is a social activity, particularly so for certain groups of people such as the elderly and mothers with young children. It can also be a leisure activity as people have increased leisure time. The social and economic elements of retailing are inter-related, as highlighted by the chairman of the Distributive Trades EDC in her Foreword to the report, 'The future of the high street':

A decrease in the economic importance of a High Street has considerable social implications. Visiting a gradually deteriorating and derelict High Street is not an attractive proposition for most customers, particularly if many of the goods they want to buy are not available. High Streets of this type may have lost their function as meeting places because few people have an incentive to visit them; even fewer linger for social purposes.

(Distributive Trades EDC, 1988: iii)

The revised PPG6 (June 1996) says very little about the social impact of new shopping developments except in seeking to ensure that development is located where all consumers are able to benefit, not just car-borne shoppers. It emphasises accessibility by a choice of means of transport, and promotes good town centre management which includes improvements to meet the needs of disabled and elderly people and those with young children.

Environmental impact

The BDP/OXIRM study identifies two types of environmental impact: *transport impacts*, in terms of traffic volumes, car parking, the provision of pedestrian areas and public transport systems, and *impacts on the built environment*, such as the change in the land use of the distribution of centres, the condition of buildings, redevelopment strategies, enhancement strategies and the level of care and maintenance (BDP/OXIRM, 1992: 79).

Major out-of-town retail developments, particularly superstores, are seen by their critics as having several (mainly environmental) disadvantages:

- extra travel time and cost entailed for customers to reach them
- their inaccessibility for some people, especially the poorest, who are forced to depend on corner shops with their limited choice and higher prices
- the severe environmental impacts of the stores themselves and their large car parks
- the environmental costs associated with energy consumption, air pollution, noise and congestion arising from the generated car mileage (Plowden and Hillman, 1995).

Out-of-town developments are attractive to many retailers and shoppers because of the availability of free car parking and good road access. At the same time, impacts relating to traffic generation, congestion and pollution in town centres are becoming much more significant. In response, initiatives are being developed which seek to limit the use of the car, e.g. park and ride schemes, traffic calming and pedestrianisation. Other forms of environmental improvement are also being used, such as the use of traditional materials and designs for paving and street furniture. Another response of the local authorities to the environmental effects of changes in retailing is town centre management, which is an attempt to replicate some of the advantages of planned shopping centres in the whole of a town centre, e.g. cleaning and maintenance, and promotion and marketing to create an image and climate of success (Guy, 1994b). The RTPI's study (1988) also concluded that environmental improvements will be necessary in town centres if they are to secure continuing commercial investment in their retail facilities.

Growing concern with the environmental impact of retail developments is reflected in government policy towards retailing. PPG6, Annex C, states that effective management and promotion of town centres will help to enhance their vitality and viability, and that good town centre management initiatives can generate civic pride among local residents and give confidence to investors and retailers.

PPG6 also states that excessive traffic and parking problems can seriously affect the attractiveness of town centres and the government therefore looks to the local authorities to promote effective management of traffic demand.

PPG13 on transport is set explicitly in the context of the government's sustainable development strategy to reduce the need to travel, influence the rate of traffic growth and reduce the environmental impacts of transport overall. It seeks to promote the vitality and viability of existing centres which are more likely to offer a choice of access, particularly for those without the use of a private car (DoE, 1994).

Retail development can also be subject to environmental impact assessment (EIA). Under the Assessment of Environmental Effects Regulations, Schedule 2, revised in 1999, an EIA may be required for urban development projects including shopping centres where the site area of the scheme exceeds 0.5 hectare. Whether an EIA is needed is at the discretion of the local planning authority, but it is unlikely that a relatively small scheme would require an EIA unless it is in a particularly sensitive location, such as a central area redevelopment in a historic town centre. The need for an EIA in respect of proposals for major out-of-town shopping schemes also depends on the sensitivity of the particular location. The new regulations are more strict than the earlier (1988) version which is reflected in PPG6. PPG6 (para. 4.19) states that for out-of-town developments a gross floor area threshold of about 20,000 square metres (gross) provides an indication of significance and the need for an EIA. It also states that, for new retail proposals in urban areas on land that has not been previously intensively developed, a development of more than 10,000 square metres (gross) may require an EIA.

Why focus on economic impact?

All aspects of impact can be important in assessing the overall effects of a development. In this book, however, attention is focused on economic impact because retail impact is fundamentally an economic concept, concerned with the diversion of trade from an existing shopping centre to a new development. Impact is generally measured in terms of percentage trade diversion from existing centres and residual turnover in these centres.

The Distributive Trades EDC (1988) report emphasised the 'health of the high street' and referred to the 'robustness' of high street shopping provision. There has been a growing concern in recent years with the vitality and viability of shopping centres which is discussed in Chapter 3. An economic focus is important both because of the economic benefits arising from new development and because of the possible adverse consequences of new development on other centres. EIA is concerned particularly with the strengths and weaknesses of shopping centres and the effects of changes in shopping patterns on the trading position of existing centres. Employment is a specialist topic in itself and involves a different methodology to the assessment of retail impact.

One of the most important but difficult elements of the application of RIA is the link between impact and loss of vitality and viability of a town centre. This link was explored by the Distributive Trades EDC's 1988 study in six steps, as follows:

- *The proposed development will have an impact on some existing businesses.* Investment in retailing which creates new and attractive shopping facilities will almost certainly affect some other traders within the same locality.
- *Impact damages businesses.* The extent to which impact may cause harm will depend on the amount of turnover lost and the profitability of the shops affected, among other considerations. A 10 per cent reduction in turnover, often cited in the past as constituting harmful impact, might force at least some shops on the margin of viability to close.
- *Damage will cause lasting harm.* The increased efficiency of new businesses may mean that new investment replaces older shops and there may be lasting harm on existing businesses.
- *Harm will be on a sufficient scale to cause extensive closure of shops.* The relative attraction of a new centre may in some circumstances lead to shop closures in smaller, nearby centres.
- *Closure of shops will not lead to redevelopment.* Closures resulting from impact are likely to be greater in less prosperous areas which may have less potential for redevelopment.
- *The absence of major redevelopment will lead to a loss of vitality and viability of a town centre as a whole.*

Demonstrating that impact goes beyond matters of competition between retailers requires an analysis of how the retail role of a town centre could be seriously affected (Distributive Trades EDC, 1988: 75–79).

The loss of even a modest amount of trade in a small centre can lead to a 'cycle of decline'. The process of such a cycle operates through an initial reduction in pedestrian flows, leading to a lower turnover level for traders. This prevents traders from upgrading their properties, creating a run-down appearance which then acts as a further disincentive for shoppers to visit the centre. Over time, this depresses turnover further and can lead to vacancies.

The importance of retail impact assessment

The importance of RIA in the British planning system lies in its application in the context of planning policy and decision-making. Impact issues are a key factor in decisions on proposed retail developments and can often be the overriding factor. This was particularly so under the policy regime established by the July 1993 version of PPG6 which stressed the need for a balance in providing for retail development between town centres and out-of-centre retail facilities. Retail impact factors remain important in the latest government guidance (DoE, 1996).

PPG6 recognises that it is not the role of the planning system to restrict competition, preserve existing commercial interests or to prevent innovation (para. 1.1). Questions of impact do arise, however, when proposals for major retail development would have an effect on the vitality and viability of a town centre. Current government policy seeks to promote town centres and restrict out-of-centre development. This policy advice is implemented by the local authorities through the development plan system and development control. Structure plan and local plan shopping policy must take the impact of new developments into account, usually through criteria-based policies on major out-of-centre developments. PPG6, Annex B, lays down guidance on the preparation of development plans, stating that the local authorities should make clear how they will assess the impact of proposals on the vitality and viability of existing town centres (DoE, 1996, Annex B, para. 4).

This book does not examine in detail the transport and travel issues involved in impact assessment because of the focus on economic factors but it shows in Chapter 6 that transport issues are becoming increasingly relevant considerations in assessing the impact of proposed retail developments. Local planning authorities are usually faced with retail impact issues in having to make decisions on planning applications. It is at this stage of the process that most impact assessments are carried out, usually by consultants acting for retailers or developers, but increasingly for local authorities confronted by proposals for development. Many local plans now have policies requiring applicants to submit RIAs with their proposals. However, local authorities may not have the expertise to understand these impact statements fully and consultants are often asked by the local authorities to provide independent audits of applicants' impact assessments.

Use of retail impact assessment in decision-making

Much of the debate on retail impact issues has taken place at public inquiries when appeals have been lodged against refusal of planning permission or when major applications have been called in for decision by the Secretary of State. There have been criticisms of the approaches used in terms of data and assumptions, and the interpretation of findings. Predictive impact assessments are especially important in deciding on planning applications. The methodology used has to be sufficiently sound to support a decision by the local planning authority. For an application to be approved there must be confidence that no significant impacts are likely on existing shopping centres. More importantly, if retail impact is to be used as a reason for refusal of an application, the basis for refusal must be able to stand up to close scrutiny at an inquiry.

It was a common experience in the early hypermarket and superstore inquiries of the 1970s to find that the significance attached by the inspector to issues of impact did not match that assumed by the parties involved.

'Whether this is because the impact arguments put before the Inspector are too complex, or too difficult to resolve, or in fact considered relatively unimportant, is difficult to determine' (Breheny et al., 1981: 472).

Holt points out that a hypothetical trade diversion figure is an inherent problem for the decision-making process. In most cases no clear picture emerges as to whether a proposed development would have a damaging effect on a particular town centre in PPG6 terms such as to warrant refusal.

> Retail impact assessment has not yet reached a level of sophistication to reduce the likelihood of widely divergent trade diversion estimates being produced by the parties at a public inquiry, leaving many an Inspector to thread his way through the disparate evidence to arrive at a reasonable assessment of the impact position.
>
> (Holt, 1998: 131-040)

The scepticism with which inspectors regard RIAs at inquiries is reflected in the July 1993 version of PPG6. It states:

> All applications for major retail developments should be supported by evidence of their likely economic and other impacts on other retail locations . . . and a broad assessment of the likely changes in travel patterns over the catchment areas. However, in assessing these factors, it should rarely be necessary to attempt detailed calculations or forecasts of retail growth or of changes in the geographical distribution of retailing. Even small variations in assumptions about trend in turnover, population, expenditure and the efficiency of use of existing retail floorspace can lead to a wide range of forecasts; and the preparation and consideration of assessments can add significantly to the cost and duration of the planning process without necessarily improving the eventual planning decision. Where the parties to a planning application or appeal are to prepare assessments, they should adopt a broad approach and seek to agree data, where possible, and present information on areas of dispute in a succinct and comparable form.
>
> (DoE, 1993a, para. 42)

The revised PPG6 (June 1996) puts this more concisely in saying that:

> Impact assessments need usually only adopt a broad approach. Parties should, where possible, agree data (such as trends in turnover, population, expenditure and efficiency in the use of retail floorspace) before preparing impact assessments and present information on areas of dispute in a succinct and comparable form.
>
> (DoE, 1996, para. 4.14)

Noel has examined the practical use of RIAs, particularly their accuracy and usefulness to land use planning. He argues that impact assessments are only useful when they accurately predict the likely impact and when they provide effective input into the decision-making process.

His case study showed that the quantitative impact assessments carried out were of limited use to the inspector in coming to his conclusions. Expert judgement and actual experience were of more value (Noel, 1989 and 1990).

The RTPI's Retail Planning Working Party examined appeal decisions on proposals for superstores in the 1980s. Among the factors that emerged at that time were that the Secretary of State generally sought to allow new shopping proposals and did not appear to have been unduly concerned about shopping impact and that he was not generally concerned at low levels of impact but, on some occasions, was concerned about the loss of an anchor supermarket (RTPI, 1988).

The policy background, of course, has changed significantly since the 1980s. The 1993 guidance states that:

> When drafting plan policies or considering planning applications for developments outside town centres, local planning authorities should take account of the possible impact (including the cumulative impact with other recent or proposed retail developments) on the vitality and viability of any nearby town centre as a whole.
>
> (DoE, 1993a, para. 33)

The 1996 guidance sets out 'key tests' for assessing retail developments: the impact on the vitality and viability of existing centres, the accessibility by a choice of means of transport and the impact on travel and car use. The 'impact test' refers to: 'likely economic impacts on town centres, local centres and villages, including consideration of the cumulative effects of recently completed developments and outstanding planning permissions' (DoE, 1996, para. 4.13).

PPG6 recognises that different types of retail development can have different impacts on the vitality and viability of town centres. For instance, the impact of a retail warehouse park selling bulky goods which cannot be accommodated easily in town centres will depend on the range of comparison goods sold. It is acknowledged that large foodstores and supermarkets often play a vital role as anchor stores in maintaining the quality and range of shopping in smaller towns and district centres.

> In the case of many smaller centres, particularly historic towns, the best solution may be the edge-of-centre foodstore with parking facilities, which enables car-borne shoppers to walk into the centre for their other business in town, and shoppers who arrive in the centre by other means of transport to walk to the store. One trip can

thus serve several purposes, and the new shop is likely to help the economic strength of the existing town centre, be accessible to people without cars, and overall generate less car use. Town centre and edge-of-centre stores facilitate a higher proportion of linked trips.

(DoE, 1996, para. 3.13)

The House of Commons Select Committee on the Environment considered RIA and said:

There cannot be a simple formula for calculating the impact of every retail development in every part of the country. However, we believe that there is a need for better guidance on anticipated impacts – especially of the long-term social and environmental effects – in retail planning, so that these lessons can be used by local authorities as they finalise their Development Plans and in their assessment of planning applications. We recommend that the Department commissions more independent research into the impacts of retail developments, particularly into the cumulative effects of out-of-town developments on the vitality and viability of existing centres.

(House of Commons, 1994, para. 112)

The government agrees that there is a need for a greater awareness of the range of impacts of out-of-centre developments, not just trade diversion. It emphasises the need for a broader assessment framework covering economic, social and environmental considerations (DoE, 1995a).

Historical development

Evolution of approaches to retail impact assessment

The development of RIA in Britain is a relatively recent phenomenon which originates from concerns about the impact of development proposals for out-of-town or out-of-centre shopping developments (Drivers Jonas, 1992: 3). Holt summarises the overall context for this development since the 1970s:

Retailing in the UK has gone through nothing short of a major revolution in the past 20 years in terms of the dramatic shift away from traditional High Street outlets to large out-of-town superstores. This fundamental change has largely been achieved despite the fact that most local planning authorities have been unhappy with it, and has been allowed to happen by central government policy . . . yielding to market pressures.

(Holt, 1998: 130-000)

The reference to 'superstores' in this statement applies to all types of out-of-centre retailing – including retail warehouses and regional shopping centres.

The historical perspective which forms the focus of this chapter covers the period from the 1960s and early 1970s onwards. An analysis is made of the approaches to RIA in each decade. However, a number of key factors apply across the whole period. Retail innovations in the 1960s led to two very important changes in the established patterns of shopping centres and shopping behaviour: increased competition between stores loosened the locational ties of stores, and economies of scale in retailing led to the development of much larger stores. There followed the growth of hypermarkets and superstores, particularly in the 1970s, and discount retail warehouses (Guy, 1994b).

Guy also comments that the history of post-war change in British urban retailing may be seen as the outcome of two conflicting influences: pressure for decentralisation, suburban development and freestanding stores from major retail firms, anxious to serve the expanding suburban market and improve efficiency and competitiveness, and pressure from local authority planners and councillors to maintain the pre-eminent position of town and district shopping centres (Guy, 1994b).

Research by BDP Planning and OXIRM (1992: 32) for the DoE notes that the first systematic studies of out-of-town retailing appeared in the early 1960s. Their report states that 'it is important to place subsequent studies in the context of a substantial historical legacy of impact research'.

The BDP/OXIRM report considers this historical context in three broad periods, each characterised by different emphases in terms of research, reflecting changing political, social and economic pressures.

> Economic impact assessment has moved from elaborate spatial modelling techniques used in the 1960s and potential impact studies in the 1970s, to a waning interest in detailed impact studies and consequently a lack of knowledge in the 1980s of the broader economic effects of new forms of retailing. There is little substantive research that can demonstrate an adverse impact of large food stores. As yet little has been done about the impact assessment of large non-food stores or regional centres. Some post-hoc monitoring exercises are under way.
>
> (BDP/OXIRM, 1992: 6)

The three time periods identified by BDP/OXIRM are the 1960s, the 1970s and the 1980s. *The 1960s* were typified by the widespread use and modification of spatial interaction modelling techniques imported from the USA. Such approaches were incorporated into the regional and sub-regional planning studies that took place during the mid-/late 1960s and they began to attract the interest of the local authorities.

In *the 1970s* there was extensive discussion about the nature of impact and the benefits or disbenefits of new stores in the context of an increasing number of proposals for out-of-centre retail development. The 1970s began with a new system of structure plans and local plans, each with specific retail development proposals. Several large-scale surveys were conducted, and particular attention was paid to the potential impact of major city centre developments.

In the context of a significant increase in the amount of out-of-town retail development, *the 1980s* were epitomised by a lack of comprehensive local knowledge on retail change, a declining interest in the impact of existing shopping centre developments, and evidence of a limited impact arising from retail development in peripheral locations.

These three time periods provide a convenient framework for the analysis of the evolution of impact assessment approaches presented in this chapter. However, a fourth time period is also relevant to this analysis – *the 1990s*. The 1990s were not highlighted in the BDP/OXIRM research but the report refers to the dramatic changes which occurred in the retail development process in Britain in the late 1980s and early 1990s, reflecting the 'shifting fortunes of the economy' and a 'swing in the pendulum of planning attitudes' towards major retail development. These recent trends have been referred to by BDP/OXIRM as a shift towards a more 'balanced approach' to retail development. The characteristics of these four decades in the historical development of RIA are examined in detail in the following sections of this chapter . . .

1960s: a decade of model development

The initial theoretical development of models took place in the USA in the 1950s and early 1960s, a period of increasing quantification in approaches to retailing problems. Gravity models were developed using a probabilistic notion of competition between shopping centres. Huff's probability model simulated consumer behaviour in making shopping trips. It is formulated as a series of probabilities of consumers choosing to visit one centre from a set of competing centres. The retail sales potential model was developed in the USA by Lakshmanan and Hansen as a derivation from earlier interaction models, enlarging upon the work of Huff. The model states that the sales potential of a shopping centre is directly related to its size, its proximity to consumers and its distance from competing shopping facilities. Lakshmanan and Hansen first applied their model in the Baltimore metropolitan area to predict the actual sales volume that would be realised in various major shopping centres given alternative planning policies and it was extensively used in studies in many parts of the USA (Davies, 1976).

The new modelling techniques, particularly spatial interaction models, were imported into Britain during the 1960s. The Centre for Environmental Studies (CES) stimulated much of the early interest in modelling. CES carried out a review of retail location models, outlining their structure and giving a selection of their applications (Cordey-Hayes, 1968).

The Distributive Trades EDC, established by the Labour government in the mid-1960s, set up a shopping capacity sub-committee in 1966 to advise on future shopping in Britain over the next 10 to 20 years. Their final report, entitled 'The future pattern of shopping', made a series of recommendations for planning policy. It was recognised that at the time there were very few out-of-town shopping developments in Britain and that on the whole central and local government held an adverse attitude to them. The report considered that it was 'highly desirable that the DoE should lay down guidelines stating criteria for their acceptability; such guidance would be of great help to all those concerned with developing a high standard of shopping provision' (Distributive Trades EDC, 1971).

As a result of this growing interest in future shopping provision and the role of models, the Distributive Trades EDC also set up a models working group which published 'Urban models for shopping studies' in 1970. This report examined the nature and purpose of shopping models, the problems they were being used to tackle and the reliability of particular models (Distributive Trades EDC, 1970). The report attempted to identify the most useful models for use in Britain based on the constraints of optimum resource allocation, economic efficiency and customer convenience (Kirby, 1986). No particular model or approach was recommended. Rather, the various types of model were found to complement each other. A step-by-step approach was suggested, with these models being incorporated into the planning process to generate and test alternative scenarios (Distributive Trades EDC, 1970).

One of the earliest British applications of the retail sales potential model was reported in work carried out by McLoughlin and others in the Department of Town and Country Planning at Manchester University on regional shopping centres in north-west England (University of Manchester, 1964). The report describes the use of the model in predicting shopping sales throughout the region. Research was carried out as a follow-up to the study noted earlier which investigated the impact of the proposed out-of-town regional shopping centre at Haydock Park. The model was seen to be very flexible in its predictive use, enabling a large number of alternative assumptions to be considered about population distribution, road networks and shopping strategies.

Various shopping models were incorporated into other major regional and sub-regional studies to provide a strategic framework for retail investment such as the Nottingham and Derby sub-regional study, the Leicester and Leicestershire sub-regional study, and the Teesside survey and plan. The retail sales potential model was also applied by Rhodes and Whitaker to retail trading in the London Borough of Lewisham to test its ability to predict the actual 1961 retail sales in ten centres of different sizes within the borough. The model performed quite successfully (Rhodes and Whitaker, 1967). Although not an impact study, this was an important early application of quantitative methods in urban planning in Britain.

Several impact studies were conducted in the 1960s by local authorities, planning agencies or bodies employed by these groups, in relation to proposals for out-of-town regional shopping centres. They focused principally on trading deflections from established centres although there was peripheral reference to questions of the quality of the environment and traffic considerations. All of the regional shopping centres proposed in the 1960s were rejected at Inquiries. The forecast losses of trade in impacted town and city centres were judged to be too high (BDP/OXIRM, 1992).

According to BDP/OXIRM there were perhaps two lasting legacies of the 1960s. First, the over-reliance on models to assess impact led to a later backlash against them. Second, the rejections of proposed out-of-town development helped to engender a 'climate of opposition' amongst planners to all types of outlying development and an assumption that the debate about major out-of-town development had been concluded.

Concern began to be expressed in the late 1960s over the methods of assessing the impact of large new stores. There was widespread opposition to increasing quantification in planning and the use of mathematical models. Critics regarded the obsession with quantitative methods as reflecting a desire to impart an aura of scientific respectability to what was regarded previously as a subjective or intuitive process. It was further claimed that the state of model development at that time was essentially theoretical, highly experimental and sporadic in its occurrence, depending very much on personal initiative and bearing little relation to planning practice.

The 1970s: a decade of debate

Interest in modelling continued in the early 1970s. Spatial interaction models were widely used to forecast the potential sales of new retail developments and to assess their competitive effects on surrounding centres. The Planning Research Applications Group (PRAG) was active in the practical development of shopping models and their application in empirical situations. The Unit for Retail Planning Information (URPI) also developed a shopping model, from a similar theoretical basis as that of PRAG. The URPI model was a combination of two approaches, embedding a spatial interaction model within a hierarchical (central place) structure. It allowed for a two-level hierarchy of higher order (comparison) goods and lower order (convenience) goods. The best known application is by South Yorkshire County Council in the preparation of shopping policies for the South Yorkshire structure plan, where the model was used for analysis of the 1974 pattern of shopping and to forecast the effects of structure plan policies in the future (Alty et al., 1979).

Throughout the 1970s there was a lively debate on the nature of impact and the benefits or disbenefits of hypermarkets and superstores. Local authorities were on the whole opposed to the growth of superstores, reflecting their 'vested interest' in established town and city centres. Most early structure plans were 'totally opposed to any interference with the established hierarchy of conventional shopping centres, were silent on the question of out-of-town shopping centres and advocated the unchallenged role of the town centre' (Holt, 1998: 131-010).

In the light of the growing number of proposals for hypermarkets and superstores, the government issued Development Control Policy Note 13 (DCPN13), on 'large new stores', in 1972, encouraging the local authorities to exercise care in granting planning permission for out-of-town development. Concern about the impact of these large new stores also led to several major surveys being carried out.

One of the first true hypermarkets in Britain was the Carrefour at Caerphilly, South Wales. This development attracted considerable research interest. Donaldsons (1979) produced three reports during the 1970s on the impact of the store after opening, after two years of operation and after five years of operation. These studies found no adverse effects on existing shops, but it was difficult to differentiate between changes resulting from the hypermarket and changes owing to other factors. Donaldsons' conclusions were that the survey findings broadly confirmed the findings of other research studies into the operations and impact of hypermarket or superstore developments. The final report states that these findings refute the argument that hypermarkets or superstores have a widespread adverse impact upon the traditional shopping hierarchy. There is no cause for concern that hypermarket or superstore developments involve unacceptable risks to established shopping areas (Donaldsons, 1979).

A comparative study of the competitive effects of three large food stores in York, Northampton and Cambridge was carried out by the Retail Outlets Research Unit at the Manchester Business School in 1976. These surveys represented, together with the DoE's survey of the Eastleigh Carrefour hypermarket, the first direct studies of the impact of superstores and hypermarkets on the trade of other retailers, as opposed to indirect studies which depended on information obtained from shoppers about their changing habits. The superstores used as case studies were Asda at Huntington, York, Tesco at Weston Favell, Northampton, and Sainsbury's in Cambridge.

Surveys were conducted of retailers before the superstores opened and approximately six months after opening. The major conclusions were that:

• The superstores had a significantly larger impact on multiple and co-operative food retailers than independent traders. The impact was greatest amongst the larger branches of these multiples and co-operatives, and smallest amongst those branches furthest from the superstores.

23

- The effect on independent grocers and specialist food retailers, like butchers and greengrocers, was very limited (Thorpe et al., 1976).

Other research was conducted on the Asda superstore in York at Huntington and, at the same time, by the Centre for Urban and Regional Research at Manchester University, on behalf of Asda. Large-scale household surveys were carried out before the store opened in 1974 and a year after it started trading. Although suburban centres close to the new store experienced some loss of trade, the effect of the store was spread over an extensive area and a large number of centres. There was no evidence that any shopping facilities, particularly in the smaller centres, were no longer viable (Bridges, 1976).

Another significant study in the 1970s was that of the Morrison Street hypermarket in Glasgow. This store, which opened in 1977, was the first such store to be built in Scotland and was also the first free-standing hypermarket in Britain to be located on an 'in-town' as opposed to out-of-town site. Based on interview surveys with hypermarket customers, the study found that the competitive impact was felt mainly by suburban supermarkets, particularly those outside the established district shopping centres (Pacione, 1979).

Mainly because of the findings of these and other studies, a less restrictive stance to proposals for large stores was taken by the government and some local authorities in the second half of the decade. DCPN13 was revised in 1977 to convey a more flexible approach to the granting of planning permissions (BDP/OXIRM, 1992: 40). It stated that proposals for large new stores would need to be studied carefully against the pattern of established shopping centres in the area, taking account of their adequacy, convenience and the need to retain their vitality and bearing in mind the planning objectives for the whole area likely to be served by the proposed store (quoted in Distributive Trades EDC, 1988: 67).

The other significant area of debate in the 1970s concerned the effects of regional shopping centres. Reference was made earlier to the Brent Cross centre which opened in 1976. A 'before and after' study of Brent Cross by the Greater London Council found 'no evidence of any other centre having been dramatically affected as a result of the establishment of Brent Cross' (GLC, 1980). By 1978 the effects of the new centre were spread widely so that no individual centre was severely affected. This was regarded by the study as confirming that the centre had filled a 'gap' in the pattern of retail outlets in this sector of London.

The late 1970s also saw the first major impact study of a city centre shopping development – the Eldon Square centre in Newcastle-upon-Tyne, which also opened in 1976. Bennison and Davies (1980) carried out research into the effects of the Eldon Square development in the context of a comprehensive account of the development, characteristics and general effects of town centre schemes. There was an initial adverse impact in terms of trade

24

losses in the more peripheral parts of the city centre but with some recovery of trade later because of an overall increase in consumer spending within the city centre. The wider regional impact was felt most acutely by Gateshead town centre which was already in decline. The findings on both the local and regional impact of Eldon Square suggest that new shopping schemes in city centres have far less adverse consequences on existing patterns of trade than is the case with similar-sized developments located on the outskirts of towns.

The 1980s: a decade of uncertainty

Since the early 1980s attention in the retail planning field has focused on several issues in relation to the impact of major shopping developments. These include: the role of hypermarkets and superstores in inner city renewal, employment issues in hypermarkets and superstores, and the effects in terms of traffic generation (Kirby, 1986).

Although the 1980s are referred to in this section as a 'decade of uncertainty', Norris (1992) has also termed this period as 'a decade of dynamic change, disarray and confusion' arising from the decentralisation of retailing and its impact on the high street. The research report by BDP and OXIRM (1992) points out that surprisingly little is actually known about the effects of major new retail developments in this period 'on the ground'. There are three reasons for this degree of uncertainty:

- There is a 'lack of comprehensive, consistent and timely knowledge on retail change'. The problems of this woeful lack of information about retail trade are discussed in Chapter 5.
- There is a 'lag effect' in appreciating the impact of major new retail development upon town centres. Interest in the effects of a new development tend to be greatest immediately after it has opened, but it will take time for these effects to be noticed and measured.
- Relating events and trends directly to the effects of major new retail development is problematic because of the sheer pace of retail change and the wide variety of other factors which affect shopping activity (BDP/OXIRM, 1992).

Retail impact studies carried out in the 1980s were set in the context of a buoyant, prosperous and expanding retail sector. Negative economic impacts were offset by general economic growth (BDP/OXIRM, 1992). This point is recognised by Howard and Davies (1993) in their research on the Metro Centre which notes that the impact of the new centre was cushioned by retail sales growth during its development period. In fact, the 1980s were exceptional in post-war times for retail sales growth.

The fostering of rapid retail growth in the 1980s coincided with a significant relaxation of constraints on retail development under the Conservative government. Davies and Howard (1988) have documented the response of the retail market to the government's *laissez-faire* approach. First of all, the new attitudes towards development accelerated the decentralisation process, leading to a proliferation of superstores and bulky goods stores, proposals for new large outlying centres, proposals for retail warehouse parks and sub-regional centres.

At the same time concerns were voiced about traditional shopping centres, some of which were beginning to show signs of 'failing health'. The impact of new outlying centres was felt to be a contributory factor in this decline. Local authorities and professional bodies such as the Royal Town Planning Institute (RTPI) became increasingly critical of the government's approach.

The RTPI's Retail Planning Working Party produced a report in 1988: 'Planning for shopping into the 21st century'. It recognised that the relative buoyancy of retail expenditure over recent years, together with increased car ownership, the improved road network, the removal of bulk food shopping from town centres and the market power of a small group of retail companies, had created significant retail opportunities for development companies. These changes gave rise to concern about the ability of planning policies to control the location of such developments. Conflicts with planning policies which had traditionally been rigid meant that a high proportion of applications needed to be decided on appeal (RTPI, 1988).

Against this background, in 1985 the Distributive Trades EDC initiated a study of the prospects for high street retailing to examine the factors affecting recent and prospective changes in shopping and to consider their longer-term implications. The study report, 'The future of the high street', highlights the uncertainty which was evident in the 1980s. The absence of clear retail policy is noted and the report comments on the major problems associated with retail data and the methodology of RIA (Distributive Trades EDC, 1988).

In the mid-1980s there were few grounds for local government opposition to the principle of superstores and, apart from technical objections, the only realistic basis on which it was possible to reject them was their adverse impact on an existing retail pattern.

> Such an examination is not an entirely exact science in the complex and volatile shopping scenarios that present themselves in most urban areas, and because clear and sustainable reasons are needed to reject planning applications the balance of decision-making went, more often than not, to the retail developer.
>
> (Holt, 1998: 131-010)

A special issue of *Housing and Planning Review* in June 1987 on shopping stated that current government policy provided little guidance on how local

authorities should deal with pressures for large out-of-town retail developments. Most local authorities were very concerned about the effect of such
developments on their own traditional town centres (Lavery, 1987). In the
same publication the pressure for large-scale retailing in the absence of
any national policy or guidance was termed 'megastore madness' (Johnston,
1987).

However, from the mid-1980s government thinking began to show signs
of changing because of the effect of the continuing growth of out-of-centre
retail developments on the health of traditional centres. In 1986 the Secretary of State for the Environment issued Circular 21/86 which introduced
the concept that large-scale retail developments should be assessed on the
basis of whether they could 'seriously affect the vitality and viability of a
nearby town centre as a whole'. Subsequently government policy was stated
in the form of PPG6, 'Major retail development', in 1988 and DCPN13 was
withdrawn. The original version of PPG6 reiterated the Secretary of State for
the Environment's statement in July 1985 setting out the general principles
of policy for new large retail developments (DoE, 1988).

The publication of PPG6 coincided with the height of a period of unprecedented pressure from large retail developments and was seen as filling a
policy vacuum. The 1988 guidance was criticised as being little more than a
consolidation of ad hoc policy that emerged at a time when there was scant
evidence of the effect of proposals for major retail developments (Roebuck
and Goddard, 1993).

The 1990s: towards a balanced approach

According to the BDP/OXIRM report the late 1980s and early 1990s were
remarkable for the dramatic changes which occurred in the retail development process in Britain. The report states:

> The unprecedented wave of new out-of-town developments in the
> late 1980s, encompassing new retail parks, large regional shopping
> centres and individual stores . . . has been followed in the early 1990s
> by one of the deepest and most protracted property crises since the
> Second World War.
>
> (BDP/OXIRM, 1992: 154)

This contrast between economic growth and a buoyant retail sector in the
1980s and economic recession and a slowdown in retail growth in the early
1990s was accompanied by changes in government attitudes towards retail
policy and the emergence of new types of retailing. The publication of
PPG6 in 1988, together with the results of a series of inquiries into out-of-
town shopping centres, had the effect of reducing much of the speculative
pressure for development. The need for a more 'balanced' approach to retail

development was recognised. This concept of balance was seen by many to be necessary to safeguard established centres and, at the same time, allow new investment. For instance, the final conclusion of research on the Metro Centre was that 'there is a challenge to find a balance between new and old forms of shopping centres, which satisfy conflicting community and consumer demands' (Howard and Davies, 1993: 149). Norris' research on RIA also recognised the need for common sense and pragmatism in retail planning, reflecting the changing attitudes of the early 1990s (Norris, 1992).

In July 1993 the DoE issued a revised version of PPG6, replacing the guidance given in January 1988. It stressed the need for a suitable balance in providing for retail development between town centre and out-of-centre retail facilities, taking account of factors such as accessibility and effective competition between retailers that will benefit consumers generally. The need to strike a balance in retail development led to the House of Commons Select Committee on the Environment report, 'Shopping centres and their future', in October 1994. In its evidence to the committee the DoE stated that the revised policies of PPG6 'do not represent a radical departure from previous planning policies but rather a rebalancing of priorities which recognise the importance of town centres: economically, socially and environmentally' (House of Commons, 1994: xx).

The Environment Committee report is very relevant to the approaches that will be taken to RIA in future years. The report notes that 'there seems to be much anecdotal but little empirical evidence of the impact of the vast majority of retail developments' (House of Commons, 1994: xliv). The DoE's evidence to the committee acknowledges that there seems to be a general lack of research on retail impact, particularly into the impacts of cumulative out-of-town development on the vitality and viability of existing centres. The report recommended better guidance on the anticipated impacts of retail developments. Regarding superstores the committee recommended that more detailed guidance be issued to local authorities on the criteria and methods to be employed in carrying out impact studies into proposed developments. Regarding retail warehouses it recommended that restrictions are placed on retail parks to prevent town centre comparison goods being sold out-of-town and that the advice given on this matter in PPG6 be strengthened.

The government's response to 'Shopping centres and their future' agrees that there is a need for a greater awareness of the range of impacts of out-of-centre developments, including trade diversion (DoE, 1995a). A project to assess the impact of superstores on smaller centres was commissioned in 1995 and was published in 1998 as 'The impact of large foodstores on market towns and district centres' (CB Hillier Parker, 1998). PPG6 was revised again in June 1996, updating and replacing the earlier guidance.

In a useful review of the future for town centres, Moss and Fellows (1995) present research findings on retail planning decisions affecting different types of retail developments in the early 1990s. The main points made at that time were as follows:

- *Superstores* – Government policy has made it very difficult for new superstores to gain planning permissions in out-of-town locations and interest from food retailers is now focusing on town centres. Discount food stores are making significant inroads into local shopping provision.
- *Retail warehouses* – Retail warehouses, particularly those selling bulky goods, are generally no longer considered a threat to town centre shopping. However there are concerns about very large, specialist out-of-town non-food developments and the cumulative effects of a number of out-of-town retailers.
- *Factory outlets centres* – Studies so far suggest that the impact of factory outlets on existing centres is likely to be minimal, but the evidence is based largely on experience in the USA which may turn out to be inappropriate in Britain.
- *Warehouse clubs* – Retail studies are limited but suggest that trade diversion from town centres is unlikely to be significant.
- *Regional centres* – Pressure for regional centres, which posed a major threat to town centres, has eased in the 1990s and government policy is likely to resist further developments of this type.

Summary

Retail impact is relevant to urban planning because of the need to assess the effects of new or proposed retail developments on existing shopping centres. Impact assessment should be set within a policy framework, therefore it is particularly important to consider how the impact of proposed retail developments should be assessed in order to make policy decisions.

Three types of retail impact may be identified – economic, social and environmental. *Economic* factors have become critical in assessing proposals for new development. *Social* impact is relevant to the extent that some groups in the population are relatively disadvantaged and less able to benefit from new types of shopping development. There is also growing concern with *environmental* impact, mainly related to the traffic implications of retail development, and this is reflected in government policy towards retailing. Attention in this book is focused on economic impact because retail impact is fundamentally an economic concept, concerned particularly with the strengths and weaknesses of shopping centres and the effects of changes in shopping patterns on the trading position of existing centres.

The importance of RIA in the British planning system lies in its application in the context of planning policy and decision-making. RIAs are carried out mainly in relation to planning applications or public inquiries, and impact questions are a key factor in decisions on proposed retail developments, particularly under the policy regime established by PPG6. The application of RIAs has, in the past, been viewed with some scepticism by planning inspectors at inquiries, and the approaches used have been refined more recently to increase their acceptability as a basis for decision-making.

The 'impact test' in PPG6 states that local planning authorities should consider the impact of proposed retail developments on the vitality and viability of existing town, district or local centres, and any cumulative effects of recent and committed developments.

Approaches to RIA have developed since the 1960s and early 1970s. The 1960s can be described as a 'decade of model development' when shopping models originally devised in the USA were applied in Britain, particularly in the context of sub-regional planning studies. Interest in modelling continued into the 1970s which was a 'decade of debate' about the pros and cons of out-of-centre retail development. Several post hoc studies of superstore developments were carried out.

The 1980s was a 'decade of uncertainty' about the impact of large new retail developments at a time of rapid retail growth and a relaxation of government policy constraints on retail development. From the mid-1980s, however, government policy began to change because of the effect of the continuing growth of out-of-centre retail developments on traditional shopping centres. PPG6 was introduced to enable the impact of large-scale retail development to be assessed in relation to the vitality and viability of existing centres.

Economic recession and a slowdown in retail growth in the early 1990s were accompanied by changes in government attitudes towards retail policy and the emergence of new types of retailing. The need for a more pragmatic approach was recognised to safeguard established centres but, at the same time, allow new investment. PPG6 (July 1993) stressed the need for a suitable balance in providing for retail development between town centres and out-of-centre retail facilities.

The latest version of PPG6 (June 1996) places the emphasis of government policy firmly on sustaining and enhancing the vitality and viability of town centres, and ensuring that new retail development is concentrated in town centres or edge-of-centre locations. Advice on assessing the impact of proposed retail developments continues to stress economic impact but, in line with the government's sustainable development strategy, it also requires assessments to be made of accessibility and impact on travel and car use. The implications are that proposals for major retail development will have to be assessed more thoroughly and that strong arguments will be needed to justify new developments which are out-of-centre.

3

POLICY CONTEXT

Approaches to RIA have evolved in parallel with the development of planning theory over the past 30 years. The concern with the relationship between theory and policy is important because, as the ideology of planning theory has developed, different stages of planning thought have produced different responses in terms of planning practice. Although it is not the purpose of this chapter to present a comprehensive review of planning theory, a brief outline is necessary first to give an overall perspective to the relationship between theory, policy and practice.

Evolution of planning theory and retail planning policy

Retailing in the context of planning theory

Four 'traditions' of planning theory have been put forward as models of the role of planning:

- *Planning as 'social reform'*. This tradition focuses on the role of the state in societal guidance and regards planning as 'scientific endeavour' (Friedmann, 1987). It led to quantitative approaches such as models for urban and regional analysis.
- *Planning as 'policy analysis'*. Here the planner's role is that of policy analyst, drawing on 'procedural planning theory' and its developments. Policy analysts are 'social engineers'. Based on systems theory, policy analysis is focused on decisions. This was a characteristically American concept and only rose to prominence in Britain in the 1970s, as the 'rational decision model' (Healey, 1991a).
- *Planning as 'social learning'*. This tradition focuses on overcoming the contradictions between theory and practice. Knowledge is derived from experience and validated in practice. Existing understanding (theory) is enriched with lessons drawn from experience and the 'new' understanding is then applied in the continuing process of action and

change. Social learning involves people-centred, community approaches (Friedmann, 1987). The planner has the role of intermediator, liaising with different agencies and trying to reconcile the conflicting aspirations of different groups in society (Healey, 1991a).

- *Planning as 'social mobilisation'.* From the viewpoint of Marxist ideology, planning here is seen as a form of politics which asserts the primacy of direct collective action 'from below'. People who have no social power of their own can expect to bring about change only when they act collectively (Friedmann, 1987). This tradition originates from Utopianism and the socialist movement which influenced early planning thought. But then the role of the planner in social mobilisation becomes inseparable from that of the politician (Healey, 1991a).

Traditional origins of physical planning

Until the 1960s planning was dominated by the physical design professions of architecture, engineering and surveying. The idea of planning theory was imported in the 1960s from North America (Hague, 1991). Before this time:

> Town planning was seen as being concerned with an attempt to formulate the principles that should guide society in creating a civilised physical background for human life. It rapidly came to be seen solely as a question of land use, layout and physical design.
>
> (Willis, 1980: 1)

Batty (1985: 104) noted that 'planning problems were treated as design problems, and design problems were largely problems of physical form dominated by questions of efficiency and aesthetics'.

The urban design model assumed planning was carried out by technical experts, with the planner acting as urban development manager. From this early perspective of the nature of physical planning, Foley (1973) suggested three propositions of the main ideologies of British town planning:

- Town planning's main task is to reconcile competing claims for the use of limited land so as to provide a consistent, balanced and orderly arrangement of land uses.
- Town planning's central function is to provide a good (or better) physical environment for the promotion of a healthy and civilised life.
- Town planning, as part of a broader social perspective, is responsible for providing the physical basis for better community life.

POLICY CONTEXT

Rational planning

By the early 1970s planning had:

> changed direction from a concern purely with the physical environ-
> ment and towards intentionally rational, comprehensive planning;
> away from a primarily practice-orientated profession towards greater
> reliance on theoretical understanding; and away from the domin-
> ation of planning by architects and engineers towards the social
> science disciplines.
>
> (Faludi, 1973)

Faludi defined planning as the application of scientific method to policy-
making. He distinguished between *normative theory*, which is concerned
with how planners ought to proceed rationally, and *positive* or *behavioural
approaches*, which focus more on the limitations of rational action.

According to Batty (1985: 100), 'urban planning . . . finally threw off its
craft image in the early 1960s and embraced the new formal rationality
through a systems approach'. Batty has termed this movement 'social engineering', reflecting a realisation that urban planning should consider social
and economic processes as well as just physical issues. He commented that:

> The urban theory considered important by planners was largely
> positive knowledge, in contrast to the planning process which was
> regarded as a normative activity. In short, it was never felt that
> planning even as a technical activity was subject to the same politi-
> cal and social pressures as other forms of decision-making.
>
> (Batty, 1985: 106)

The purpose of planning as a comprehensive rational process was to help
society and individuals to achieve their goals. Planning work focused on the
processes and methods by which goals were translated into objectives and
alternative courses of action defined. Some people linked planning as a rational decision process with the idea of regions and urban areas as integrated
systems of social and economic activity. The objective of planning then
became the design of 'control systems' to guide urban and regional change.

The development of what came to be known as procedural planning
theory, rooted in general systems theory, became a vogue in the late 1960s
and early 1970s (Hague, 1991). The promotion of the systems approach
towards rational planning was perhaps best advocated by McLoughlin who
saw the need for a fundamental reorientation in both the conceptual basis
and the practical operations of planning. The position at the start of the
1960s was one of rapid evolution in the development of theories of human
locational behaviour. Emphasis was shifting rapidly away from earlier static

equilibrium notions which made little attempt to offer behavioural explanations (McLoughlin, 1969). McLoughlin stated that:

> Planning seeks to regulate or control the activity of individuals and groups in such a way as to minimise the bad effects which may arise and to promote better 'performance' of the physical environment in accordance with a set of broad aims and more specific objectives set out in a plan.
>
> (McLoughlin, 1969: 59)

In the rational decision model the planning process becomes a series of steps or phases in a cycle, as follows:

- the formulation of goals and objectives
- the identification of possible courses of action
- the evaluation of alternatives
- the making of decisions
- the implementation of action, and
- the review of performance.

Neo-Marxist ideology

Neo-Marxist planning theory developed in the 1970s. The neo-Marxist viewpoint criticised the available literature on urban planning theory on the grounds that urban planning was treated as an abstract analytical concept and that it was normative, problem-solving and idealist/Utopian. The urban planning process was regarded as a social and historical phenomenon within urbanised capitalised society (Scott and Roweis, 1977). From the Marxist perspective the role of planning in contemporary society can only be understood by recognising the structure of modern capitalism as it relates to the physical environment. It is argued that the fundamental social and economic institutions of capitalist society serve the interests of capital at the expense of the rest of society. Public ownership and centralised planning would replace existing market and political decision processes. Marxists have been highly critical of traditional planning theory and practice. Planners' attempts to employ scientific techniques and professional expertise are seen as helping to legitimise state action in the interests of capital by casting it in terms of the public interest, neutral professionalism, and scientific rationality (Klosterman, 1985).

Free market ideology

At the start of the 1980s there was a reaction against rational decision-making. The rise of the New Right gave free market ideas a new currency

(Hague, 1991). Planning in the 1980s was subject to increasing centralisation of planning powers and the encouragement of private enterprise. Sorensen and Day (1981) pointed out that, in the social sciences, and particularly economics, there has always been an intellectual right-wing tradition stressing freedom of the individual, the benefits of both market forces and entrepreneurship, the role of law, and the perils of bureaucratic control of the economy and society. Free market planning argues that the market is in the best position to judge what society wants. The planner would only intervene to express the public interest. This view envisages a shift in perspective involving a greater emphasis on free choice and market determination in contrast to the increasing levels of planning control characteristic of the 1960s and 1970s. It implies a significant reduction in government control including detailed planning control.

This free market ideology is critical of normative theory on the basis that planning, unlike other professions, is not a natural component of a market economy; it is largely a creation of government, and political power and bureaucracy cannot ensure optimal welfare for society.

The ideology maintains that public sector decision-making must generally be impoverished when compared to market mechanisms (Sorensen, 1983). Free market ideology argues that the market system is inherently a better method for satisfying human wants and aspirations than recourse to government. The intellectual basis for this doctrine comes from market theory and public choice. Public choice theory represents the application of economic methodology to the study of politics (Self, 1993). Public choice theorists model the study of politics on the methods and assumptions of neo-classical market economics. Originally developed in the USA, the basic assumptions of public choice theory are of self-interest and rationality. The ideology seeks to limit state intervention and liberate market forces, e.g. through privatisation. In Britain these ideas were strongly pushed by the Thatcher government. Public choice theory has contributed to attempts to control bureaucracy. In Britain in the 1980s, for instance, Conservative governments mounted successive assaults upon the traditional role of local government.

Public choice theory has had an influence on policy towards retailing in Britain, as shown by the *laissez-faire* attitude of the early to mid-1980s. The free market ideology of the 1980s has now run its course and the new vogue will be for mechanisms which operate in a market framework but with a stronger measure of public accountability (Hague, 1991). However, free market thinking has not disappeared entirely. A 1996 paper from the Institute of Economic Affairs argues against further regulation of retailing which is an efficient and innovative industry. The argument is that, if people want to shop for bulky items in more convenient locations than the traditional high street, they should be able to do so wherever possible.

The central policy question of the moment concerns the use of the planning process to limit out-of-town or edge-of-town development to situations where the would-be retailer is able to prove that the new superstore will have no impact on local shopping centres. Such a rule favours the existing edge-of-town retailers . . . and the larger multiple retailers who, unlike their smaller competitors, have the resources to hire the necessary experts to challenge planning decisions.

(Burke and Shackleton, 1996: 89)

Pragmatic approaches

The route along which planning theory developed in the 1990s has shown a divergence of views and it appears that no clear conceptual framework has yet been proposed to replace the procedural planning theory of the systems approach. A number of ways forward have been suggested. Healey (1992a), for instance, advocates what she terms 'communicative rationality' or 'planning by debate' using principles of logic and scientifically formulated empirical knowledge to guide actions. Breheny and Hooper (1985) state that the recent history of public policy-making, particularly planning, shows a distinct move away from supposedly 'rational' approaches towards a reliance upon more pragmatic procedures. They refer to the 'troubled theory–practice debate in planning since the 1960s':

To many planning theorists the hope has been that the disappearance of the procedural model would mean a move from a mechanistic or repressive mode of planning to something much more enlightened, open and progressive. We might expect, then, that the movement away from this technical model might make planning less 'controversial' and less 'mundane'. Unfortunately the opposite appears to have happened, at least in Britain. Planning has become increasingly pragmatic, less open to scrutiny, more conservative, and more susceptible to 'private government'.

(Breheny and Hooper, 1985: 14)

Breheny has elaborated on the widening gap between planning theorists and practitioners. His view is that the theorists have been excessively abstract in their work, have ignored planning practice and have failed to offer any prescriptive advice to practitioners. Practitioners, on the other hand, have continued to lack any sustained critical assessment of their own activities and have lapsed into an 'insular pragmatism'. The result has been theory which has become increasingly useless and practice which has become increasingly devoid of intellectual credibility (Breheny, 1983).

36

The radical changes which have occurred both in planning theory and practice since the 1970s involve a combination of two themes, according to Batty. First:

> Planning theory has undergone a revolution from a concern for process to one for product, from a concern for means to one for ends, from a concern for positive knowledge to one for normative knowledge, and from a concern for realism to one for idealism.

and second:

> Changes in planning practice where a concern for comprehensive strategic thinking has given way to pragmatic short-term responses. In this, formal rationality, which in its clearest expression was embodied in the systems approach, has all but disappeared.
>
> (Batty, 1985: 118)

Intervention of planners in the retail market

This section focuses on the implications of the theoretical context for retail planning and particularly for policy decisions on retail development.

> Land use planning is concerned with government intervention in the private land-development process. The purpose of this intervention is to achieve particular social, economic and physical outcomes by the control of land development. Decisions, usually in the form of policies, are implemented by the process of development control.
>
> (Breheny, 1983: 106)

Guy considers that there are three reasons for planners to intervene in the retail market:

- to improve the efficiency of its operation
- to act when the operation of an uncontrolled market is likely to lead to inequities in the level of service to the local population
- to allow new investment or environmental improvements, or negative, e.g. traffic, visual intrusion, or competitive impact.

The planner's role is seen as regulating development, largely by prohibiting or modifying proposals which would cause harm to interests of acknowledged importance and accepting proposals that do not cause such harm (Guy, 1994b). According to Kivell and Shaw (1980), intervention in the retail market is generally justified by planning authorities on one or more of the following grounds:

- The retail case – it is desirable to restrain free market forces in order to prevent an excessive number of shops and in order to promote an optimum mix and range in any location.
- The urban case – the arrangement and location of shopping facilities exerts a strong influence upon other facilities and upon urban form in general.
- The social planning case – in the distribution of shopping facilities, as with other resources, planners have taken responsibility for ensuring that all sections of the community are adequately served.
- The environment case – the planner endeavours to separate, or reconcile, non-conforming land use activities and minimise the adverse environmental impact of new development (Kivell and Shaw, 1980).

These arguments clearly reflect the main components of the dominant ideology of the planning profession, i.e. the search for order in the urban system and for balance between land uses and between retail centres, with the aim of serving the 'public interest'. Planners have traditionally been discouraged from taking social and economic factors into consideration in formulating policies and making development control decisions. But social and economic issues are central to retail planning since retailing is a form of social service and also contributes substantially to the local economy of every town or city (Gibbs, 1987). Gibbs believes that the scope of planning is particularly confused in relation to retailing. She states that:

> The prevailing attitude of the planning profession towards retailing is inherently conservative. Planners, and local authorities in general, are philosophically and financially committed to a centralised, hierarchical structure of retailing; a commitment which is further reinforced by the statute.
>
> (Gibbs, 1987: 15)

The position has shifted since the 1980s, however, through changes in government policy. In the 1990s there was a drawing back from the extremes of market-led ideology.

> There would appear to be a consensus that the development of land should continue to be controlled through the medium of plans prepared by democratically elected bodies within the framework of broad policy guidelines set by local government, with the teeth of such a system being the power to refuse planning permission for development subject to appeal to a higher authority.
>
> (Holt, 1998: 020-095)

Theoretical issues and concepts in retail development

Relevance of theory to retail planning

Planning policy towards retailing has been determined by political and economic changes, notably:

- political resistance to new forms of shopping in the 1970s
- rapid economic growth in the 1980s and a relaxation in government policy towards out-of-centre development
- economic recession in the 1990s and new controls over retail development outside town centres.

RIA developed out of the new ideas of procedural planning theory and social engineering in the 1960s. Like planning theory itself, approaches have changed over the last 40 years. RIA is essentially an aid to decision-making and, therefore, the rational decision model remains particularly relevant. But approaches to RIA have become less theoretical and more pragmatic in response to the requirements of the planning system in Britain. For the purposes of advising decision-makers on development proposals, the planning process expects decisions to be soundly based and justified.

Neo-Marxist ideology has never been a significant influence on retail planning because it is fundamentally opposed to the capitalist system which underlies the retail development process. During the 1980s free market ideology briefly influenced retail planning in relaxing government policy towards out-of-centre development but, on the whole, issues of retail impact imply a degree of government intervention over the operation of the retail market.

In the 1990s, planning took a more pragmatic approach. Government policy since 1996, enshrined in PPG6, seeks to regulate the location of new retail development. It restrains free market forces and introduces environmental as well as economic factors into decisions on major shopping development. RIA is concerned primarily with the economic effects of shopping development, derived from the rational decision model, but with increasing emphasis being placed on environmental impacts in the public interest.

Influence of theoretical concepts on retail policy

Retail development is demand-led; it will not take place without a retailer or developer who seeks to promote new development. In the retail property market the private sector plays the dominant role as supplier. Retail development also requires the involvement of planners and ultimately the support of the local authorities in obtaining planning permission. The planning system is no longer conceived of as being opposed to the market, and planning authorities need to respond to the challenge of 'market-sensitive' planning (Healey, 1992b).

Therefore, in the evolution of planning policy, the interaction between the private sector and the regulatory authorities is critically important. The relationship between 'state and market' in planning in the last decade has been examined by Healey. Her conclusion is that:

> The planning system has a high degree of adaptability. This allowed the substantial shift to a narrow agenda in the early 1980s, to a negotiated project-based practice dominated by development-market values in the later 1980s, and to strategic concerns for the integration of economic, environmental and social issues in the arena of managing land-use and environmental change in the early 1990s.
>
> (Healey, 1992b: 430)

The theoretical basis of retail planning was initially founded on central place concepts and the retail hierarchy. Gibbs (1987) comments that planners' philosophical commitment to the retail hierarchy has been reinforced by the support of the local authorities towards town centre redevelopment schemes and the preservation of the established pattern of shopping. She says 'by protecting and enhancing the role of the town centre in this way, local authorities have encouraged the development of a centralised, hierarchical structure of retailing' (Gibbs, 1987: 9).

Gayler (1989) is critical of the defence of the town centre and its position at the top of the retail hierarchy by the British planning system since the 1950s. He claims that retail innovation has been seen as a threat to the existing order. Rees (1987) recognises that retail trends have led to a marked erosion of the hierarchical arrangement of centres which planners have traditionally sought to defend. Planners have done this for several reasons:

- to maximise 'social and territorial equity' in the spatial distribution of shopping opportunities
- to maximise the economic benefits of functional integration
- to assist in co-ordinating the ancillary, publicly-financed infrastructure and services in the most cost-effective manner.

In attempting to reconcile centrality and dispersal, planners have tended to be torn between entrenched protectionism and reluctant acceptance of those new forms of decentralised retailing which offer identifiable advantages (Rees, 1987).

Shepherd and Thomas (1980) comment that central place theory provides at best only a partial explanation for intra-urban shopping behaviour. Kivell and Shaw (1980) also regard central place theory as being over-simplified and having severe limitations. Potter (1982) made an evaluation of theories and models of urban retail location and concluded that central place theory and its derivatives remain of great significance to retailing studies. He argued that concepts such as range, threshold and hierarchy are still valid and have an influence on the way the planning system intervenes in the retail market.

Batty (1997) has commented on the way in which the retail revolution represents a move to increased productivity through technological change. He has said that the traditional hierarchy of shopping centres will disappear entirely because of technological change, mail order and teleshopping. However, at the present time, retail hierarchies continue to have a significant role in structure plan and local plan shopping policies. The revised PPG6 shows the continuing importance attached to the concept of a hierarchy of centres in government policy. It states that structure and local plans should set out the hierarchy of centres and the strategy for the location of shopping and other uses, and:

> In particular, the development plan should indicate a range and hierarchy of centres, from city centre, through town centre, district centre to local centres and village centres, where investment in new retail and other development will be promoted and existing provision enhanced.
>
> (DoE, 1996, para. 1.5)

General interaction theory has not had such a strong influence on retail policy as central place theory, but it underlies the formulation of shopping models which have been used in RIA. The most widely used hierarchical shopping models, developed by PRAG and URPI in the 1970s, were based on an integration of central place theory and a spatial interaction framework which was found to give the most accurate representation of shopping trips (see Chapter 4).

The spatial transformation of retailing that has occurred with decentralisation and new forms of shopping development has tended to break down the traditional hierarchical structure. The policy response to these trends was initially to resist change, then to accommodate it until pressure for decentralisation became out of control and, finally, to seek a balance between town centre and out-of-centre provision. Guy notes that the debates over off-centre development have perhaps obscured some more general issues concerning planning intervention in the retail system.

> Changes in planners' control over retailing have reflected wider changes in government attitudes towards property development. In the 1970s, government interests supported the view that planners should determine the broad location and type of development, in

the interests of protecting existing (generally unplanned) facilities and providing good quality shopping for the local population. In the 1980s the government moved to a position of leaving the impetus for retail growth and change to the private sector developers and retailers. Local authorities were no longer expected to specify the location and type of new development. However, leaving these decisions to the market led to intense pressure for off-centre retailing, leading to the boom in regional centre proposals in the mid-1980s. Faced with this pressure, and the possible consequences for established town centres, the government has now offered more support for town centres and has attempted to set more stringent conditions for off-centre development, in the revised version of Planning Policy Guidance Note Six.

(Guy, 1994b: 201)

Current issues in retail planning

Key issues

Many issues currently face the planning system in dealing with retailing in Britain. Three general issues can be identified which currently represent a major challenge to the planning system:

- How can the planning system keep pace with retail trends?
- How can the decline of town centres be reversed?
- What are the implications of government policy on sustainability for retail development and town centres?

How can the planning system keep pace with retail trends?

Changes in retailing occur faster than the planning system can respond. Planning policy tends to follow behind. The major pressure for change in retailing has been decentralisation. Three waves of retail decentralisation occurred in Britain during the 1970s and 1980s: the development of major food stores outside town centres, the growth of retail warehouses selling bulky non-food goods, and major out-of-town regional shopping centres selling mostly comparison goods and competing directly with town centres (Schiller, 1986). A fourth wave is now also emerging in the form of factory outlet shopping centres, also mostly in out-of-town locations.

Other factors are also important, such as socio-economic trends (rising car ownership, suburbanisation of the population, growth in consumer expenditure and changing lifestyles) and changes in the retail industry itself. Pressures for out-of-centre development seem likely to continue despite current government policy.

How can the decline of town centres be reversed?

Some town centres and district centres are in decline. Some have experienced a serious loss of trade because of new developments and some have failed to attract investment. It is often thought that town centre decline or poor performance is due to excessive competition from out-of-town retailing. That is true in a few cases, such as Dudley or Sheffield, but is not generally so (Baldock, 1996). The view is widely held that some older city centres had been in decline long before the arrival of out-of-town competition. Out-of-town centres may have accelerated change, but they did not initiate it. Town centres have generally failed to provide the quality of shopping environment demanded by consumers.

As part of the research study by the Urban and Economic Development Group (URBED), 'Vital and viable town centres: meeting the challenge' (1994), an extensive postal survey was carried out of the opinions and judgements of local planning authorities in England and Wales on the current state and immediate future of town centres. In response to a question about the overall health of town centres, the vast majority said they were vibrant, improving or stable. Only a fifth said they were in decline. It may be then that the current concern over the apparent decline of Britain's town centres has been overstated. However, the results did reveal that specific types of town have been more severely affected by decline than others. These are the older industrial and commercial towns, and suburban centres (Simmie and Sutcliffe, 1994).

Whether they are in decline or not, it is certainly the case that town centres face pressures. Changes in planning guidance have come too late for many small businesses and specialist food retailers who have suffered from the growth of superstores in particular. As well as the continuing decline in the number of small retailers, there are pressures resulting from the trend towards retail services (dry cleaners, pharmacies, post offices, etc.) in large supermarkets. The effects of decentralisation can be seen but it also has to be accepted that there has been a lack of attention given to the positive promotion of town centres. Town centres need to be promoted and managed if they are to compete effectively with out-of-town centres. Government policy has moved towards the need to *promote*, not just protect, the vitality and viability of existing town and suburban centres, through both the development plan process and through decisions on individual planning applications.

What are the implications of government policy on sustainability for retail development and town centres?

Concerns arose in the 1980s over the increasing consumption of natural resources and pollution of the natural environment. This has led to a concern for 'sustainable development', that is development which does not lead to the depletion of non-renewable energy sources. The aim of sustainable development is to search for a path of economic progress which does not impair the welfare of future generations. Related to these concerns is the idea that urban development should be designed so as to reduce the need for car travel (Guy, 1994b). Government policy on sustainable development was set out in the previous version of PPG1 which stated:

> The Government has made clear its intention to work towards ensuring that development and growth are sustainable. It will continue to develop policies consistent with the concept of sustainable development. The planning system, and the preparation of development plans in particular, can contribute to the objectives of ensuring that development and growth are sustainable. The sum total of decisions in the planning field, as elsewhere, should not deny future generations the best of today's environment.
>
> (DoE, 1992a: 1)

The current PPG1 (February 1997) continues to emphasise the contribution of the planning system to achieving sustainable development and PPG6 seeks to achieve sustainability in retail development. The government's latest strategy on sustainable development, 'A better quality of life', defines sustainable development as 'ensuring a better quality of life for everyone, now and for generations to come' (DETR, 1999b).

Decentralisation of retailing

Decentralisation of retailing has occurred over a period of 40 years through the development of supermarkets, superstores, retail warehouses and parks, and purpose-built out-of-town centres. Most food shopping and bulky goods shopping is now done outside town centres. The major movement out-of-town has been largely driven by developers and not retailers, apart from in the food and bulky goods sector. Certainly, the regional shopping centres have typically been developer-driven, and at weak points in the planning system. The Metro Centre and Merry Hill, significantly, were built in Enterprise Zones (Pope, 1996).

The dominant urban trend in post-war Britain according to Breheny (1993) has been the ongoing decentralisation of population and jobs away from the larger cities to small towns and villages. Some commentators regard this as counter-urbanisation, i.e. not simply a process of extended suburbanisation but a definite rejection of urban living. The trend towards decentralisation of retailing is part of the same pattern of social and economic change. It is probably unrealistic to expect that the high street will once again become the exclusive retail heartland of our towns and cities. It is better to accept change and work with it. The effect of decentralisation may be that town centres will increasingly serve opposite ends of the market – upmarket speciality shops for the affluent, and convenience/discount shopping for the elderly, poor and those on benefit who cannot afford to travel.

Chase and Drummond (1993) have examined the growing threat to town centre shopping arising from decentralisation. They note that many British cities have developed more along North American rather than European lines, even though our cultural heritage is essentially European. The decline of the town centre is seen as part of a broader, long-term trend towards an increasingly suburban lifestyle. They present two alternative planning scenarios for the millennium:

- *Laissez-faire – the suburban society*. The retailing industry will lead the way to cheaper quality shopping out-of-town and the long-term decline of town centres.
- *Intervene and invest*. A national planning policy linked with properly resourced town centre authorities will create a climate for confidence in town centres. Retailers and investors will continue to dominate but they will have a much wider spread of interest in all town centres.

Planners and developers are urged to look away from the North American scenario to the European model, with a more interventionist approach to retail planning (Chase and Drummond, 1993).

Decentralisation will no doubt remain a major issue in retail planning. There are still a large number of planning consents for out-of-town developments. PPG6 cannot alter this situation other than to urge local authorities to consider the implications of new or emerging development plan policies designed to sustain and enhance town centres when existing consents for out-of-town development come up for renewal.

The plan-led approach

The planning system has been 'plan-led' since the advent of Section 54A of the Town and Country Planning Act 1990. PPG6 emphasises the need to regenerate the city, town and district centres using a 'plan-led approach' to guide development. Annex B in the guidance gives advice on development plans and sets out points to be taken into account in preparing structure plans and local plans. As well as offering this policy guidance, PPG6 recommends how local authorities should identify sites for retail development. In summary, it says:

- local plans should identify a range of suitable sites to meet demand
- local planning authorities should adopt a sequential approach to selecting sites for new retail development; the first preference is for town centre locations, then edge-of-centre

- a town centre strategy could be prepared as part of the local plan process.

The plan-led approach is now established as a policy mechanism but it will take time for plans to respond to the new guidance. In the meantime there is concern among retailers about the availability of sites for retail development. In many town centres the easy and best located sites have already been developed, and all that is left are peripheral sites. The alternative may be incremental development, such as infill, and extensions to existing shop properties and selective, often small-scale redevelopment (Baldock, 1996). For the plan-led approach to succeed, local authorities must be proactive in identifying town centre sites and those sites must be realistically capable of being implemented within the lifetime of the local plan. Unitary development plans (UDPs) and local plans must provide a clear lead in meeting future shopping needs. Local authorities need to assess the potential for new shopping development and to consider appropriate locations for inclusion in plans.

Sustainable development

The 1990s became the decade of sustainable development. The nature of 'sustainability' is complex but essentially it means that the role of planning becomes one of contributing to the reconciliation of two seemingly incompatible aspirations: to achieve economic development and higher standards of living and to protect and enhance the environment, now and for future generations (Selman, 1995).

Healey and Shaw (1993) believe that the concept of sustainable development offers a new approach to planning – the challenge of integrating and relating the economic, social and physical dimensions of human existence. But they are critical of the political will required to achieve environmental objectives. They argue that 'with respect to ideology much of the present discussion of sustainable development ideas and of plan-led planning is little more than symbolic tokenism' (Healey and Shaw, 1993: 774).

Others are also critical of approaches towards sustainable development. Tiffin says that in practical terms, the best that planners can do is to try and secure an improvement in future environments by looking for reductions in consumption through movement. In the case of retail development, we can try to steer new facilities to those places which are best served by a mix of transport modes (Tiffin, 1996). The Environment Committee observed that policies for sustainable development are being made in the absence of any knowledge about how sustainability is quantified.

Breheny has argued that to achieve sustainable development, planning policies need to have more positive environmental objectives, for example reducing pollution, recycling, reducing energy consumption, reducing travel demand and making public transport more attractive and economical. Such an approach would include refusing planning permission for new out-of-town retailing (Breheny, 1993). Certainly, new stores or shopping centres that are accessible only by car, or that are likely to stimulate more or longer car journeys, are much less likely to be approved now than would have been the case in the 1980s (Guy, 1994b). The latest version of PPG6 strengthens the government's opposition to out-of-centre development.

Given freedom of choice, shoppers will use their cars to make shopping trips when it is convenient to do so. Thorpe (1994) points out that in practical shopping terms there are four types of trip:

- the bulk food shopping trip
- the day-out (the 'let's go shopping' syndrome)
- the morning parade ('popping down to the high street')
- the purchase trip, for specific non-food items.

Of these, the bulk food shopping trip is usually car-based by necessity. The 'day-out' often involves long distances but low spending and so is likely to be environmentally unsustainable. The 'morning parade' involves only local trips and often includes both food and non-food shopping. The 'purchase trip' is highly organised, and efficient in environmental terms.

For town centre regeneration to compete against out-of-centre developments, there will be a need to make town centres more attractive by providing more or better car parking but also by having a positive incentive for shoppers to use public transport and leave the car at home. The answer is not to restrict the use of the car, but to make real investment in the alternative forms of transport. For most shoppers, the bus is still not an acceptable alternative to the car.

PPG6 says that town centres can play an important role in reducing the need to travel and to rely upon the car. Sustainability is the basis for two of the 'tests' in PPG6. As well as the key 'impact test' on the vitality and viability of existing centres, new retail developments will be judged according to their accessibility – new retail development should be located where it is accessible by a choice of means of transport (usually in or near to town centres), and impact on travel and car use – local planning authorities should assess the likely proportion of customers who would come by car and the size of the catchment area.

PPG13 is intended to help meet the commitments in the government's sustainable development strategy to reduce the need to travel, influence the rate of traffic growth, and reduce the environmental impacts of transport overall. There is currently a debate about whether the emphasis of government policy on town centres will result in less car mileage. To help provide the information upon which policy should be based, a survey project was carried out using the trip rate information computer system (TRICS) database on a sample of Safeway stores in town centre and out-of-centre locations. The results showed that although town centre stores attract a slightly lower percentage of their customers by car, town centre stores involved longer travel distances and more vehicle mileage than out-of-centre stores. It is somewhat ironic that out-of-centre stores may have a more positive effect on reducing vehicle mileage than town centre stores (Eastman, 1995).

Government policy guidance

Government policy on retailing

PPG6

Government policy on retailing in England is the subject of PPG6. The current version of PPG6 was issued in June 1996, replacing the July 1993 version which itself superseded the guidance issued in January 1988. The 1988 guidance was formulated in quite different economic circumstances than those of the late 1990s. The underlying themes in 1988 were best summed up as 'competition and choice' (Raggett, 1994a). The July 1993 version shifted the balance towards sustaining and enhancing town centres and adopted a more cautious approach towards out-of-centre retailing.

PPG6 (July 1993) made four main points:

- there should be support for competition, within a clear planning framework
- the planning system should be positive about the role of town centres
- new retail development will continue to be supported as long as it does not undermine the vitality and viability of town centres
- there should be support for town centre management and partnership.

During 1994, ministerial statements began to depart from the formal policy guidance. The balance of policy advice tended to alter towards safeguarding the vitality and viability of town centres and towards a greater emphasis on access to shopping centres by transportation modes other than the private car. Only a year after it was revised, the government announced that PPG6 was to be revised again. The need for further revision was reinforced by the decision of the House of Commons Select Committee on the Environment early in 1994 to review a wide range of retail issues. The Environment Committee's report, 'Shopping centres and their future', was published in October 1994 (House of Commons, 1994). It takes a critical view of retail trends and retail planning policies, and recognises that planning guidance has to be clear and consistent in order for retailers, local planning authorities and the development industry to operate efficiently and effectively. It accepts the need for flexibility so that planning policy can provide general guidance for application in a local context. A large number of recommendations were made regarding the changes to PPG6.

The government published its response to the Environment Committee's report in February 1995. On the matter of the revision of PPG6, the government stated that it agreed with most of the recommendations for further advice and proposed to revise PPG6 accordingly. The draft revision of PPG6 was issued in July 1995 and the final revision was published in June 1996. The revised guidance contains some significant shifts in policy emphasis.

The two main strands of current government policy can be summarised as sustainability and support for town centres. With *sustainability* there is an emphasis in PPG6 and PPG13 on reducing the use of the car for shopping trips. The guidance acknowledges that town centres can play an important role in reducing the need to travel and to rely upon the car. With *support for town centres* as shown earlier, there is now firm support for enhancing and promoting town centres. This is embodied in the introduction of the sequential approach to selecting sites for new retail development. The sequential approach is discussed in Chapter 6.

New retail developments are expected to be in accordance with the strategy for retail development set out in the development plan. The revised guidance does not preclude out-of-centre developments but sets out three key tests for judging applications which fall outside the development plan framework. The key features of the guidance are as follows:

- On *planning for town centres and retailing*, there should be emphasis on a plan-led approach to promoting development in town centres, both through policies and the identification of locations and sites for development, emphasis on the sequential approach to selecting sites for development, and support for local centres.
- On *town centres*, there should be promotion of mixed-use development and retention of town centre uses, emphasis on the importance of a coherent town centre parking strategy in maintaining urban vitality, promotion of town centre management, and good urban design.
- On *assessment of retail proposals*, there should be guidance on clarifying the three key tests for assessing retail developments (impact on vitality and viability of town centres, accessibility by a choice of means of transport, and impact on overall travel and car use), on how to assess out-of-centre developments, and on how certain new types of retail development should be assessed (DoE, 1996).

The Planning Minister, Richard Caborn, said in a speech to the RTPI Conference in June 1999:

The last government's permissive policy before 1993 did a great deal of damage to town centres, especially smaller towns and district centres. Indeed, half of the current out-of-town shopping floorspace was built in just five years between 1986 and 1991. We are only now beginning to see the benefits of applying the new policy firmly, fairly and consistently. The government has no intention of changing the policy on retail development which is directed to promoting the viability of our town and city centres.

(DETR, 1999e)

PPG13

The element of government policy relating to transport issues is dealt with in both PPG6 and PPG13. A complete revision to PPG13, 'Transport', was published in March 1994, which further developed the concept that new development should recognise a need to reduce the length and number of motorised journeys, encourage alternative means of travel and hence reduce reliance on the motor car. Retail development features specifically in PPG13, which states:

Structure plan policies for retailing should seek to promote the vitality and viability of existing urban and suburban and rural centres. Shopping should be promoted in existing centres which are more likely to offer a choice of access, particularly for those without the use of a private car.

Travel for shopping has grown strongly, particularly in the non-food sector. In local plans, authorities should:

– maintain and revitalise existing central and suburban shopping centres by enabling development to take place there and by policies which improve the quality and competitiveness of those areas;
– encourage local convenience shopping by promoting the location of facilities in local and rural centres, and ensuring such areas are attractive and readily accessible on foot or by bicycle;
– where suitable central locations are not available for larger retail development, seek edge-of-centre sites, close enough to be readily accessible by foot from the centre and which can be served by a variety of means of transport;
– avoid sporadic siting of comparison goods shopping units out of centres or along road corridors; and
– provide for both local shopping and residential uses in large new developments, where feasible.

(DoE, 1994: paras 3.9–3.10)

A draft revision of PPG13 was published in October 1999 (DETR, 1999c). The revised guidance reiterates the policies on retail and leisure uses in PPG6 and says that the overall approach by local authorities on shopping should take account of:

- the accessibility by public transport, walking and cycling
- the linking of planning and transport strategies
- the allocation of key sites for major development which are accessible by public transport
- transport assessments.

Transport assessments are particularly significant and they are referred to in detail in Chapter 6.

Therefore, through PPG6 and PPG13, government policy is attempting to restrain market forces in retailing and has introduced social and environmental as well as economic factors into decisions on major shopping development.

Government advice in Wales is contained in Technical Advice Note 4, 'Retailing and town centres', which accompanies Planning Guidance (Wales), 'Planning policy', 1996. The advice note is consistent with PPG6 and reproduces relevant extracts from the English guidance.

Policy and practice in Scotland

Government policy guidance on retailing differs in some respects in Scotland from that in England and Wales. The original Scottish Office advice in the form of national planning policy guidelines (NPPGs), 'The location of major retail development', was produced in 1986 (Scottish Development Department, 1986). A draft revised NPPG, 'Retailing', issued in February 1995, was an attempt to shift the focus of guidance away from out-of-centre developments to existing centres. It promoted a much stronger role for town centres and town centre management initiatives and provided backing for a more interventionist approach by local authorities where out-of-town retail development raises issues of sustainability, design quality, and impact on the vitality and viability of town centres (Duffill, 1995). Out-of-centre developments were not ruled out but greater weight was attached to their impact and potential to undermine the vitality and viability of existing town centres.

Most controversially, the draft NPPG went further than PPG6 in attempting to quantify 'significant' impacts on the vitality and viability of town centres from out-of-centre retail developments. It stated that significant impact may be judged as a trade diversion of 10 to 15 per cent, although the impact will vary in the light of the particular circumstances of each existing centre. No justification was given for the figures adopted and there was widespread criticism of the use of such criteria. Drysdale (1995) stated that none of the recent reports and research papers on retail impact makes any reference to particular thresholds of retail impact and some documents – notably the Scottish Office's own 1992 research paper on 'Retail impact methodologies' – explain that there can be no standard threshold of impact because of the countless variations in circumstances and in the methods of calculation.

The Scottish guidance was issued as NPPG8, 'Retailing', in April 1996, just before PPG6. Like PPG6 it incorporates the sequential approach to site selection and the emphasis on town centres. It states that:

- out-of-centre developments should not be of such a scale as to undermine the vitality and viability of town centres
- new retail development should be sited where there is a choice of transport and should not be dependent solely on access by car
- all applications for major retail developments over 5,000 square metres gross should be supported by a RIA which provides evidence of likely economic and other impacts on other retail locations (compared with 2,500 square metres in PPG6)
- where assessment shows clear evidence that a planning proposal would have a significant impact on an existing centre and would undermine its vitality and viability, permission for development should be refused.

The controversial percentage impact threshold mentioned in the draft NPPG was left out of the final version. This omission is seen as a weakness in the guidance (Braithwaite, 1997). NPPG8 and PPG6 offer similar guidance but some commentators believe that the Scottish guidance is clearer and firmer because it promotes town centres more effectively and gives greater certainty for investment (Davies, 1996).

A further review of NPPG8 was produced in October 1998. In Scotland the sequential approach classes district and local centres as 'town centres', moving them ahead of edge-of-centre locations. The Scottish guidance also requires the consideration of converting vacant and underused town centre premises as suitable locations for development.

Government advice to local authorities

PPG6 is of fundamental importance to local authorities. Local planning authorities must take the content of PPGs into account in preparing their development plans and the guidance may also be material to decisions in individual planning applications and appeals. In the revised PPG6 (June 1996) the government sets out the objectives that local planning authorities should adopt in their planning policies.

The advice to local authorities in PPG6 is specified in Annex B, 'Development plans'. Annex B refers to regional planning guidance, structure plans and local plans. On *regional guidance*, it states that regional planning guidance (and strategic planning guidance in the metropolitan areas) is concerned only with the overall strategy for a network of centres in the region, as a guide to the preparation of structure plans and UDPs, including assessing the scope for new regional centres.

On structure and local plans, Annex B makes the following key points:

Structure plans should provide a clear strategy for town centres and retail development within the county, helping to ensure a consistency of approach between districts. They should also indicate whether there is a role for retail developments outside town centres.

Local plans should generally conform with structure plans. In preparing local plans, local planning authorities should:

- take account of future retail demand
- consider the relationships between centres
- assess the effectiveness of previous local plan policies on vitality and viability
- identify a range of suitable sites for development
- include criteria-based policies
- set out policies for retention of town centre uses.

Local planning authorities are also advised that:

- the views of property owners and retailers should be taken into account in assessing the capacity of town centres to accommodate growth
- plans may distinguish between primary and secondary frontages in town centres and their relative importance to the character of the centre should be considered
- retailing policies and proposals in development plans should be based on a factual assessment of retail developments and trends, including surveys to obtain up-to-date data on existing shopping provision, and the future potential.

Because of the government's concern with the cumulative effects of major new retail development on existing centres, it issued a circular, 'Shopping development direction' in 1993. The direction required local planning authorities in England and Wales to notify the Secretary of State of proposals for shopping development in excess of 20,000 square metres gross floorspace or smaller proposals of more than 2,500 square metres gross which would exceed 20,000 square metres when aggregated with other retail developments in the previous five years within a radius of ten miles, before granting planning permission (DoE, 1993b).

Since 1992 local planning authorities in England have also been required to notify departure applications to the Secretary of State if they consisted of more than 10,000 square metres of gross retail floorspace, and if, by reason of their scale, nature, or location, they would significantly prejudice the implementation of development plan policies and proposals (DoE, 1992b).

New guidelines were issued in DETR Circular 7/99 which says that retail schemes of 5,000 square metres gross floorspace or more must be referred to the Secretary of State, compared to 10,000 square metres previously (DETR, 1999d). The new circular also applies to mixed-use commercial schemes.

These procedures provide the Secretary of State with the opportunity to call in applications for decision, usually where the proposals are of more than local importance (DoE, 1996). Specific advice is given to local planning authorities in PPG6 on how they should deal with retail policy issues in development plans and how to assess applications for new retail developments.

Vitality and viability

The phrase 'vitality and viability' was introduced by the Secretary of State for the Environment in response to a parliamentary question in July 1985. He said:

> Since commercial competition as such is not a land use considera-
> tion, the possible effect of proposed major retail developments on
> existing retailers is not in this sense a relevant factor in deciding
> planning applications and appeals. It will be necessary, however, to
> take account in exceptional circumstances of the cumulative effect
> of other recent and proposed large scale retail developments in the
> locality, and to consider whether they are on such a scale and of a
> kind that could seriously affect the vitality and viability of a nearby
> town centre as a whole — for example whether they seem likely to
> result in a significant increase in vacant properties or a marked
> reduction in the range of service the town centre provides, such as
> could lead to its general physical deterioration and to the detriment
> of its future place in the economic and social life of the community.
> Town centres need to maintain their diversity and activity if they
> are to retain their vitality, but the range and variety of shops and
> services will change, as they have always done, in response to chang-
> ing conditions.
>
> (quoted in DoE, 1988, para. 7)

Vitality and viability have been defined in the URBED report, 'Vital and viable town centres', as follows:

Vitality refers to how busy a town is at different times and in different parts.

Viability refers to the capacity of the centre to attract continuing investment, not only to maintain the fabric but also to allow for improvement and adapting to changing needs.

(URBED, 1994: 55)

Therefore vitality means liveliness and activity while viability suggests commercial survival and the continued attractiveness of a centre. How to measure vitality and viability has been problematical. In the late 1980s the Distributive Trades EDC concluded that the best data for examining the viability of different high streets related to shop vacancies associated with a lack of economic prosperity. But its report stated that vacancy rates are only a partial measure of shopping centre performance; qualitative indicators may also be important, e.g. the overall appearance of shops and other buildings (Distributive Trades EDC, 1988).

The BDP/OXIRM study for the DoE included a survey of local authorities which indicated that economic impacts in the form of physical deterioration were the most appropriate measure of lack of vitality and viability, i.e. poor environment, acute congestion and a lack of opportunity sites available to compete with out-of-town development (BDP/OXIRM, 1992).

The revised PPG6 in July 1993 gave added importance to the concept of vitality and viability. It elaborated on the concept, stating:

Good retailing contributes to the vitality and viability of town centres. But vitality and viability depend on more than retailing; they stem from the range and quality of activities in town centres, and their accessibility to people living and working in the area.

(DoE, 1993a, para. 5)

The guidance suggested some basic measures of vitality and viability, acknowledging that in practice most aspects of vitality and viability will be difficult to assess with confidence. The following indicators, it suggested, can usually provide the main criteria for the purposes of a planning application or appeal: commercial yield on non-domestic property (see below), and pedestrian flow.

Other factors which may be relevant are listed as:

• the proportion of vacant street level property in the primary retail area
• the diversity of uses
• retailer representation and profile
• retailer demand or intentions to change representation
• the physical structure of the centre (DoE, 1993a, Figure 1).

The June 1996 version of PPG6 has revised the indicators of vitality and viability. The following indicators are now recommended for assessing the health of town centres:

- the diversity of uses (shopping, offices, leisure, pubs, restaurants, housing, etc.)
- retailer representation and intentions to change representation (i.e. demand)
- shopping rents within primary shopping areas
- the proportion of vacant street level property
- the commercial yields on non-domestic property
- pedestrian flows
- accessibility
- customer views and behaviour
- the perception of safety and occurrence of crime
- the state of the town centre environmental quality (DoE, 1996, Figure 1).

The full version of Figure 1 in PPG6 is included in this chapter as Table 3.1.

Some comments are needed on the appropriateness of the indicators in Table 3.1, particularly on yields. PPG6 says that information on yields should be used with care. The main source of yield information is the Inland Revenue Valuation Office Agency which has monitored prime retail yields in 550 shopping centres throughout England since 1994 and publishes data for April and October each year. Yield is defined as:

> a measure of property value which enables values of properties of different size, location and other characteristics to be compared. It is the ratio of rental income to capital value, and is expressed in terms of the open market rents of a property as a percentage of the capital value. Thus, the higher the yield the lower the rental income is valued, and vice versa. A high yield is an indication of concern by investors that rental income might grow less rapidly and be less secure than in a property with a low yield.
>
> (Valuation Office, 1998: 95)

It is acknowledged that factors which affect yield are complex and need to be interpreted with reference to the circumstances in each individual town. Broadly speaking, however, low yields indicate that a town is considered to be attractive and as a result be more likely to attract investment than a town with high yields. The average shopping centre yield in Great Britain in 1998 was 7.6 per cent. The lowest yields were in Meadowhall, Sheffield (4.0 per cent) and in a number of affluent city centres such as Chester, Oxford,

Table 3.1 Measuring vitality and viability

The following indicators are useful for assessing the health of town centres. They provide baseline and time-series information on the health of the centre, allow comparison between centres and are useful for assessing the likely impact of out-of-centre developments. Local planning authorities should regularly collect a range of these indicators, preferably in co-operation with the private sector.

- *The diversity of uses* – how much space is in use for different functions, such as offices; shopping; other commercial, leisure, cultural and entertainment activities; pubs, cafés and restaurants; hotels; educational uses; housing, and how has that balance been changing?
- *Retailer representation and intentions to change representation* – it may be helpful to look at the existence and changes in representation, including street markets, over the past few years, and at the demand from retailers wanting to come into the town, or to change their representation in the town, or to contract or close their representation.
- *Shopping rents* – this is the pattern of movement in Zone A rents within primary shopping areas (i.e. the rental value for the first 6 metres depth of floorspace in retail units from the shop window).
- *The proportion of vacant street level property* – vacancies can arise even in the strongest town centre, and this indicator must be used with care. Vacancies in secondary frontages and changes to other uses will also be useful indicators.
- *The commercial yields on non-domestic property* (i.e. the capital value in relation to the expected market rental) – this demonstrates the confidence of investors in the long-term profitability of the centre for retail, office and other commercial developments. This indicator should be used with care.
- *Pedestrian flows* – the numbers and movement of people on the streets, in different parts of the centre at different times of the day and evening, who are available for businesses to attract into shops, restaurants or other facilities.
- *Accessibility* – this is the ease and convenience of access by a choice of means of travel, including the quality, quantity and type of car parking, the frequency and quality of public transport services, the range of customer origins served and the quality of provision for pedestrians and cyclists.
- *Customer views and behaviour* – regular surveys of customer views will help authorities in monitoring and evaluating the effectiveness of town centre improvements and in setting further priorities. Interviews in the town centre and at home should be used to establish views of both users and non-users of the centre. This could establish the degree of linked trips.
- *The perception of safety and occurrence of crime* – this should include views and information on safety and security.
- *The state of the town centre environmental quality* – this should include information on problems (such as air pollution, noise, clutter, litter and graffiti) and positive factors (such as trees, landscaping and open spaces).

York and Edinburgh (4.5 per cent), but yields can be as high as 10 to 12 per cent in poor centres.

Yields are seen as fluctuating dramatically as they are affected by a number of variables including the wider external influences of the investment market. As such they will be difficult to use to measure the change in viability through time. Stockdale (1993) argues that commercial yields have little to do with the viability of a town centre and that they are more a reflection of macro-economic circumstances. He maintains that the most compelling indicators of a town's retail health are the rental levels within the town and the performance of the retailers themselves. Other factors of relevance such as vacancy rates and retailer representation, it is claimed, are easier to collect and should have an equal status in the assessment.

Evidence presented to the House of Commons Select Committee on the Environment in 'Shopping centres and their future' by the Royal Institution of Chartered Surveyors (RICS) also casts doubt on the value of employing yields as a main indicator. In its response to the report the government agrees with the RICS that the yields used as a comparative indicator of relative attractiveness of commercial investment in a town centre need to be correctly analysed and their limitations understood (DoE, 1995b).

Several organisations have developed indicators of the health of shopping centres to assist professionals working in this field to compare centres and monitor changes over time. Three examples are described below.

Investment Property Databank (IPD) has devised a 'total return measure' which includes both changes in the capital value of property over 12 months and the rental income received by landlords. A three-year average has been selected to smooth out annual fluctuations in total returns. The data are used to rank centres in terms of their commercial performance (IPD, 1996).

The company *Experian* introduced a retail centre ranking in 1998, in the form of a national and regional ranking of 550 centres in Great Britain, based on their 'vitality score'. The vitality score is built up from seven variables – national multiples, comparison retailers, amount of floorspace occupied by multiples, vacancy rates, retail density, the presence of key retailers, and the total number of shops. The larger centres clearly come out on top. Therefore, it is a guide to the size and strength of centres, not their vitality and viability.

The *Lockwood Survey* by John Lockwood of Urban Management Initiatives is a study of 250 town and city centres in Great Britain based on an analysis of store-takings data and local factors which have affected the vitality and viability of individual centres in the 1990s, such as the quality of the shopping environment and the streetscape, car parking and access, and retail confidence. Details are given of the strengths and weaknesses of each centre. The study was first carried out in summer 1997 and updated in spring 1999 (Urban Management Initiatives, 1999).

In addition to its recommendations on indicators of vitality and viability, the URBED report (1994) considers that it is important to look at the underlying components of a healthy town centre. These can be analysed through a 'health check' which should focus on three basic qualities of the health of a town centre: *attractions* – retail provision, services, variety and number of attractions; *accessibility* – access to the centre, and movement within the centre; and *amenity* – appearance and character of the centre, and security (the three 'A's).

URBED suggests detailed appraisals of town centres to indicate how particular aspects of a centre are performing. PPG6 accepts this advice which is embodied in the key indicators in its Figure 1. However, there are differences between the URBED health check approach and the PPG6 indicators, and advice is given in Chapter 6 on how these differences can be reconciled in practice.

Tesco Stores Limited commissioned Healey and Baker (1995) to conduct a critical analysis of the approach adopted by the government in PPG6 to the measurement of vitality and viability. Healey and Baker's report examines whether or not the indicators set out in PPG6 can be regarded as appropriate tests of the vitality and viability of a shopping centre and whether such indicators are capable of independent and reliable measurement, such that they could satisfactorily be used to inform the decision-making process, either at local or national level.

The report has some significant conclusions:

- Most aspects of vitality and viability are difficult to assess with confidence. There are serious reservations about the use of commercial yield on non-domestic property – only limited reliance should be placed on commercial yields when considering the vitality and viability of a centre.
- Data relating to pedestrian flow is not capable of direct translation into shoppers and potential retail trade. Information on pedestrian flow is only likely to be useful if it has been collected on a consistent basis for a particular centre over a number of years. Even then, great care will be needed in the interpretation of the reasons for any changes.
- The proportion of vacant street level property in the primary retail area can be helpful, as long as there is a consistent basis of measurement and an understanding of general retail market conditions and specific local circumstances.
- Other indicators are regarded as being of limited value – the diversity of uses in a centre, retailer representation and profile, retailer demand, and physical structure of the centre.

Healey and Baker recommend those elements which they consider would usefully form part of a relatively simple and straightforward approach to vitality and viability. It includes both quantitative and qualitative factors, and the approach may need to be varied depending on the size of the centre, its retail function, and the purpose for which the assessment is being made. Three factors are identified:

- *The function of the centre* – this is based on Goad Trader Plans and household surveys.

- *Vacancy levels and retailers' requirements* – these are preferably based on the percentage of retail floorspace within the defined primary frontage.
- *Impact assessments* – these are to evaluate the effects of new retail proposals in out-of-centre locations.

Summary

Approaches to RIA have evolved in parallel with the evolution of planning theory over the past 40 years. Planning has changed from a concern with the physical environment and design to a concern primarily with social and economic issues. In the 1970s procedural planning theory and the systems approach formed the basis for rational planning and the development of shopping models. Other theoretical perspectives rooted in neo-Marxism and the free-market ideology in the 1980s were replaced in the 1990s by pragmatic approaches more relevant to policy in practice.

Urban planning is primarily a form of state intervention in the development process. The planning system intervenes in the retail market to regulate free-market forces in the public interest. Planning policy towards retailing has been determined by political and economic changes since the 1970s. In the 1990s planning became more pragmatic and policy-orientated, and RIA has developed in response to changes in the policy context in the 1990s. Retail planning has traditionally been based on normative spatial models of shopping behaviour, particularly central place theory and its conceptual basis of a hierarchy of shopping centres. General interaction theory forms an alternative approach to explaining shopping patterns and provides the theoretical base for shopping models. The concept of a hierarchy of centres has been weakened by the spatial transformation of retailing that has occurred with decentralisation and new forms of shopping development.

Some key issues face the planning system in keeping pace with trends in retailing, in trying to reverse the decline of some of our town centres, and in meeting the aims of the government's sustainable development strategy. Decentralisation of retailing in Britain has led to a decline in the role of town centres. The latest version of PPG6 (issued in June 1996) emphasises the need to regenerate city, town and district centres, using a plan-led approach to guide development including the identification of suitable sites, adopting a sequential approach to the selection of sites, and recommends strategies and action plans for improving and managing town centres. Policy towards sustainable development is important in aiming to reduce shopping trips by private car, and PPG13 seeks to reduce travel demand.

Government policy on retailing in England and Wales has been revised significantly since the late 1980s. Rapid economic growth in the 1980s and free-market ideology led to a relaxation in government policy towards out-of-centre development, which is reflected in the first PPG6 (1988). But the economic recession of the early 1990s and a widespread reaction against the effects of decentralisation led to new controls over retail development out-side town centres. The approach now seeks to support town centres and restrict out-of-centre development, and to reduce the use of the private car for shopping trips. PPG6 does not preclude out-of-centre development, but spells out the criteria against which proposals should be assessed. Govern-ment policy has also been revised in Scotland, in the form of a new NPPG8, 'Retailing', which is broadly in line with PPG6 but has some differences in emphasis.

PPG6 gives advice to local planning authorities on retail policy in devel-opment plans and on dealing with planning applications for new retail development. It also directs that local authorities should inform the DoE (now DETR) of proposals for major retail development which may be called in for a decision by the Secretary of State. PPG6 also offers limited advice on the use of RIA but there is still no clear guidance from the government on how RIA should be applied in practice.

It is no longer sufficient for those preparing RIAs to look only at the quantitative impact of proposed developments, examined in terms of trade diversion. Government policy now expects that the impact will also be assessed qualitatively by considering the health of those centres affected and the ability of centres to withstand trade loss. Issues of the accessibility of a new store and its impact on travel patterns are also material considerations.

The concept of vitality and viability of town centres, which forms the cornerstone of government shopping policy, is still being refined. A range of preferred indicators is recommended in PPG6 but pedestrian flows and com-mercial yield are no longer regarded as the best indicators to use. Yield is a notoriously difficult indicator to use in practice and it has been widely criticised. A broader-based approach using qualitative 'health checks' to assess vitality and viability seems to be the way forward.

4

THE CONVENTIONAL
METHODOLOGY OF RETAIL
IMPACT ASSESSMENT

The assessment of retail impact is an area of urban planning which evokes many different reactions from planners, politicians, developers and planning inspectors. Attitudes vary from enthusiastic to uninterested, confused, sceptical and even antagonistic. Apart from those who are specialists in the field, there appears to be a lack of understanding among those involved in retail development about how RIA is carried out. Unfortunately this lack of understanding extends to those who have a role in decision-making on retail development proposals, such as local authority officers and members, and inspectors.

It is fair to say that RIA is generally not held in high regard, probably because of the unsatisfactory way in which it has been used in the past as well as because of the problem of lack of understanding. It has been described as 'arcane' and a 'black art'. Howard has said:

> Impact studies have become a particular kind of necromancy; gazing into the future and using spells, or formulae, to predict the shape of things to come, without even knowing a great deal about the present condition of the subject.
>
> (Howard, 1988: 1)

A comprehensive research study on RIA methodologies was carried out by Drivers Jonas for the Scottish Office (Drivers Jonas, 1992). This study deals very clearly with the main issues inherent in the methodology. Information has also been obtained from planning consultants who are particularly active in the field of RIA, and from direct experience of undertaking a large number of RIAs.

To understand the methodology it is first necessary to know the requirements of the decision-makers. In order to obtain a realistic prediction of the likely effects of a new retail development, decision-makers usually require a thorough examination of the existing retail situation, particularly on shopping patterns, and existing shopping centres; a statistical analysis of the likely catchment area, turnover, trade draw and market share of the proposed development; and an assessment of the trade diversions expected as a result of the new development. In addition, decision-makers require RIAs to be accurate, reliable and impartial (Drivers Jonas, 1992).

These requirements are discussed in detail later in this chapter. Drivers Jonas' analysis states that decision-makers are usually concerned with two principal issues: the need or justification for a retail development, and the likely effect which the development may have on the vitality and viability of established shopping centres.

A third issue, related to the second, is that of location; whether the proposal can be regarded as 'off-centre' or 'out-of-centre', or whether it will in effect form part of an existing centre and be supportive of its vitality and viability.

Quantitative need for additional retail floorspace is often calculated by comparing shopping supply and demand in the form of existing retail floorspace (and estimated turnover) in an area and retail expenditure in the same area. An excess of expenditure, in the form of leakage of spending out of the area, is regarded as an opportunity for additional provision. Quantitative need can also be expressed as capacity for new development when demand and supply are used in forecasting future need. A growth in demand arising from population and expenditure growth up to the forecast year represents potential for new development. The current turnover estimate is subtracted from the forecast potential turnover estimate to provide an estimate of 'expenditure headroom'. Expenditure headroom is translated into 'floorspace capacity' by applying appropriate turnover to floorspace ratios. Estimates of turnover potential derived in this way are not very reliable because the actual turnover may be higher than the level estimated. Such 'capacity studies' are often treated with scepticism because they take a very simplistic view of future shopping needs.

Qualitative need is usually defined as a sectoral or geographical gap in the distribution of facilities. Less commonly, planners are concerned with a deficiency in the quality of provision. Drivers Jonas' analysis showed that reporters in Scotland deciding planning appeals usually place greater weight on subjective assessments of qualitative deficiencies in preference to calculations of demand and supply. In England and Wales there seems to be even less reliance by inspectors on quantitative assessments of need (Drivers Jonas, 1992).

It is also important to consider the location of a proposed retail development and whether it will divert trade from or bring trade into the town in which it is sited. If the latter applies, the issue of 'impact' may not be relevant because the new proposal will be supportive of the town centre rather than be in competition with it. The question of location has been given added significance by the inclusion in the revised PPG6 of the 'sequential approach'. In some circumstances it may be that conventional impact-testing is not thought necessary at all, particularly in clear edge-of-centre situations. Here local policy may permit large retail developments anyway and, even if not, it may be taken for granted that there would be overall benefit.

Overview of the methodology

A variety of approaches has been developed over the years to assess retail impact. The use of different methods depends on a number of factors, including:

- the type and scale of the development proposed, e.g. foodstore, retail warehouse, regional shopping centre or factory outlets
- the type of area, e.g. conurbation or rural area
- the availability of local data
- whether the analysis is being carried out before or after a new development has taken place.

A further factor is the client's budget for the assessment. Low resources could mean a broad-brush approach; greater resources could allow a comprehensive approach including shopper/household surveys.

Three main approaches have been developed:

- *post hoc studies* of retail performance which attempt to draw a statistical picture of the trading patterns which emerge as a result of a new shopping development
- *shopping models* which use a mathematical formulation to reproduce existing trading patterns in a spatial system and assess the effects on expenditure flows of new retail developments
- *predictive impact assessments* (or 'a priori' studies) which are used to determine the likely effects of proposed retail developments, either before a planning application is decided or to support an appeal against the refusal of planning permission for new retail development. These are undertaken manually using a step-by-step approach. PPG6 recommends that the parties adopt a broad approach to preparing assessments, and seek to agree data where possible and present information in a succinct and comparable form.

Post hoc studies

The early retail impact studies carried out in the 1970s were post hoc studies, based on surveys of the trading effects of hypermarkets and superstores. Three principal kinds of survey were carried out:

- consumer questionnaire surveys which were conducted before and after a store opened

- consumer questionnaire surveys which were conducted only after a store or centre had been opened; people were asked to recall earlier patterns of shopping behaviour
- questionnaire surveys of retailers in the vicinity of a new development which were conducted either by postal or personal interview (or both) (Davies, 1984).

There have been criticisms of these post hoc studies. Kivell and Shaw (1980) thought that the empirical approach to post hoc impact studies in the 1970s led to few firm conclusions. The studies adopted a weak methodology and were too case-specific. They recognised the need for a major review of the impact of superstores and hypermarkets on existing trade patterns, e.g. by assessing the various methodological frameworks and comparing the findings of different studies. Davies and Kirby (1980) also criticised the weak methodological base of the early research studies in Britain. For instance, they inferred that all retail changes were the direct result of the new store but, of course, wider factors must also be considered. Perhaps the major weakness, however, was that these studies did not use a stated methodology and often their findings were incompatible.

More recently post hoc studies have concentrated on the impact of the regional out-of-town shopping centres built in the 1980s. OXIRM has studied the Metro Centre and Meadowhall, while Roger Tym and Partners has studied the impact of the Merry Hill Centre. Howard cautions about using the lessons of earlier studies to apply to regional centres.

> Past practice and existing literature are not, or should not be taken as, a good guide to assessing the impact of a regional out-of-town shopping centre. Most past work has been concerned with single superstores, or perhaps a superstore plus a few smaller units.
>
> (Howard, 1986: 282)

Shopping models

Chapter 2 reviewed the historical development of shopping models in Britain during the 1960s and early 1970s. Five approaches to the development of shopping models in the 1960s were identified by Cordey-Hayes (1968), as follows:

- aggregation of individual behaviour
- central place theory
- retail gravity models
- the intervening opportunities model
- maximum entropy models.

These five approaches are not mutually exclusive and can complement each other in applications. The interdependencies between them can be arranged schematically, as shown in Figure 4.1.

The retail gravity model became the most widely used in theoretical and practical applications in Britain, particularly the version developed by Lakshmanan and Hansen in the USA, but other variations were also developed. Most retail location models were developed either in an academic context for exploratory research purposes, or in an ad hoc way, often by local authority planning departments or planning consultants (URPI, 1986).

In reviewing the use of shopping models, Guy observed that:

> Impact assessment relating to new retail developments became a major concern in British town planning at around the end of the 1960s but, as one public inquiry followed another, it became clear that apparently abstruse disagreements on matters such as zone size, choice of travel cost measure and methods of calibration could lead to radically different conclusions about the extent of the trading impact of new developments. Public inquiry Inspectors and participants became increasingly tired of such discussions. Eventually the Department of the Environment advised implicitly against the use of mathematical models in impact studies.
>
> (Guy, 1991: 191)

After the mid-1970s public sector concern over the potentially detrimental effects of new retail developments continued but methods of impact assessment reverted to 'step-by-step' analyses, which are discussed in detail in this

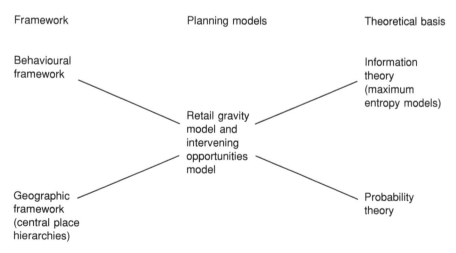

Figure 4.1 Approaches to the development of shopping models

chapter. Guy (1991) considers that the most important development in the use of shopping models has been in the private sector. Several major retail chains now use models for forecasting both the turnover of proposed new stores and the impact of these stores upon existing stores operated by the same company.

The theoretical and operational problems of the retail gravity model have been reviewed by several writers, including Jensen-Butler (1972), Davies (1976), Kivell and Shaw (1980), and Batty (1985). These problems arise from the gravity theory proposition that trade will be drawn to a centre in direct proportion to its size and in inverse proportion to distance. The main problems can be summarised as follows:

- There is a lack of a theoretical base; the model attempts to generalise about individual behaviour from aggregate empirical data.
- As an equilibrium model this model can only allocate given certain static conditions, but shopping behaviour is not static.
- Choosing the input variables for the attraction and deterrence functions – attraction can be based on retail sales or floorspace, and deterrence can be based on time or distance.
- Choosing the area units – the size, shape and number of zones, and the overall spatial limits to the study area.
- The calibration of shopping models – finding the appropriate values for the parameters.
- The prediction of the future size of centres, population, expenditure, etc.

The major problem relates to the calibration of shopping models. Research on the complexities of calibration has been undertaken by a number of academics, notably Batty and Saether (1972), Openshaw (1973), and Guy (1991). The difficulty lies in the lack of data on observed expenditure flows within the retail system. Openshaw notes that 'Generally it is possible to fit a model after a fashion to any empirical data but unless notice is taken of its statistical validity, the results may be totally meaningless' (Openshaw, 1973: 370). Openshaw concluded that no satisfactory shopping model calibration can be made when the trip pattern is not known. It is necessary, therefore, to collect shopping trip data, preferably by undertaking a diary survey of all shopping trips made by a household.

The models produced in the 1960s were very crude and were often inappropriate simplifications. Four major criticisms emerged:

- The theory was at an extremely low level.
- The data required by the theory was often unavailable, and sometimes not measurable, and this meant further arbitrary simplification when models were constructed.

- The models often posed computational problems of size and solution.
- What could be theorised about and what could then be modelled often did not match the precise requirements of the planners and policy-makers (Batty, 1985).

The early shopping models attempted to describe forces of interaction between zones and centres. They did not formally predict future shopping patterns nor capacities, but provided a quantitative analytical base from which to predict by projecting the exogenous variables. In modelling incremental growth from an existing distribution of centres, they had some value in indicating the volume of sales that could reasonably be expected. But it was recognised at the time that many theoretical and practical improvements needed to be made.

Further developments in shopping models occurred in the 1970s. The URPI developed a 'hierarchical' shopping model, *Shop*, which combined central place and spatial interaction theories within a hierarchical structure of retail sectors and trading areas. Then in the 1980s a new generation of shopping model was introduced as a logical successor to *Shop*. The *Markets* model uses a 'more realistic' method of predicting expenditure patterns and allows the user more scope to apply detailed knowledge of a local area. The user determines each centre's trade area by specifying the range of shops and shopping centres with which each zone is permitted to interact in the model. The allocation of expenditure from a particular zone is restricted to a limited number of shops and centres (URPI, 1986).

Drivers Jonas' review of shopping models comments on the disadvantage of the time taken to calibrate the *Markets* model and the need for survey data on existing shopping patterns to enable the model to be calibrated accurately. It also notes that *Markets* is not intended for use in modelling different types of trip, and that it deals only with shopping trips originating from a shopper's home location.

> For all its limitations, the *Markets* model and similar gravity-based models are a useful reminder that the turnover of existing and planned shopping centres must be related to available spending within the catchment area and also inter-related between centres. The turnover potential of one new development will be limited initially by the continued turnover demands of existing facilities and subsequently by any other new developments taking place in the system. Thus turnover estimates for new developments ought to take account of other commitments, proposals or developments under construction, as well as the total amount of money available in the catchment area.
>
> (Drivers Jonas, 1992: 77)

The *Markets* model has been repackaged by The Data Consultancy (formerly URPI) as a 'dynamic analysis tool for turnover predictions'. It is being marketed to the retail industry for use in assessing new store sites and the performance of existing stores, along with sophisticated computer mapping and graphics facilities. *Markets* remains 'a general purpose, singly-constrained gravity model, designed to model consumer shopping trips and expenditure flows over geographic space' (The Data Consultancy, 1999a).

Predictive impact assessments

As the use of shopping models has diminished since the 1970s, there has been a development of manual approaches to RIA based on a series of steps in the prediction of impact. Usually the calculations are carried out manually or with the assistance of a spreadsheet, though some consultants have devised computer software for the purpose of estimating existing expenditure flows and then calculating the trade diversion caused by a new store introduced into the system. The 'conventional market share approach' according to Guy (1987) assumes that consumer expenditure is fixed and that new stores simply lead to a redistribution of this expenditure.

Current applications of the step-by-step approach have developed from the method applied in Gloucestershire by Breheny et al. (1981). Faced with a Savacentre hypermarket inquiry at Barnwood, Gloucester, in 1979, 'a method was devised which attempted to combine the realism of the pragmatic approaches and the supposed rigour of the modelling efforts' (Breheny et al., 1981: 461).

> Essentially the procedure was based around a set of steps assembled from the guidelines in DCPN13, and the approaches adopted by other local authorities, plus a certain amount of innovation. The steps were:
>
> (1) determine future levels of available expenditure per capita
> (2) apply estimates of expenditure to forecast population in each isochrone; this gives total available expenditure
> (3) determine likely turnover of store
> (4) determine proportion of turnover coming from customers living in set isochrones from the store
> (5) for each isochrone, to take the share of turnover coming from that isochrone as a percentage of total available expenditure in that isochrone
> (6) determine where the expenditure 'captured' by the new store would be spent alternatively; this gives an assessment of the impact on existing or planned centres.
>
> (Breheny et al., 1981: 464)

69

The sequential nature of the approach is shown by the authors in Figure 4.2.

Of the six sequential steps in Figure 4.2, two were regarded as being of crucial importance: the calculation of store turnover and the impact of the proposal on existing centres (Roberts, 1982).

In the last decade the step-by-step approach has been further refined and different consultants have developed their own variations of the approach. An outline of the main steps in the methodology is presented below.

Steps in the methodology

Drivers Jonas' review lists the main steps or stages in the approach to RIA which represent a 'common methodology', as follows:

- identify the catchment or study area
- estimate the expenditure within the catchment area
- estimate the turnover of existing shopping centres
- estimate the turnover of the new retail proposal
- estimate the amount of spending in each existing centre which will be diverted to make up the new store's turnover, and the locational source of that spending
- express the amount of diverted trade from each shopping centre as a percentage of the estimated pre-impact turnover of that centre.

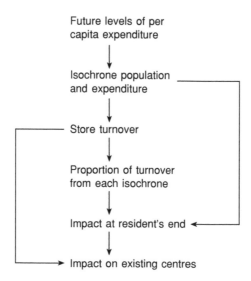

Figure 4.2 The step-by-step approach to retail impact assessment

Very few methods do not involve all of these stages, although variations in approach to some of the stages mean that the 'running order' may vary. Inevitably the disagreements as to how to calculate impact generally arise towards the end of the process, usually from stage 4 onwards (Drivers Jonas, 1992: 60).

Simplifying the 'traditional' step-by-step approach into its key elements, the following main steps can be identified in the methodology of RIA:

Step 1: definition of the catchment area of the proposed development

It is conventional to define the primary catchment area of the proposed development, that is the area from which the store or centre will draw the majority of its trade. Ideally information should be obtained from a household survey to show existing shopping patterns in the area, but often data on shopping patterns are not available. The catchment area may be subdivided into drive time isochrones, and isochrones may be further subdivided into zones for greater spatial accuracy in assessing impact.

Step 2: expenditure estimates

Existing expenditure in the catchment area, or the spending power of residents, is derived from population and per capita expenditure in the base year projected to a design year, usually one to two years after the expected opening date of the new development. If isochrones or zones are used, generated expenditure is produced for each isochrone/zone. Expenditure is divided into convenience and comparison sectors, either by goods type or business type.

Step 3: turnover of existing centres

Information must be obtained on existing shopping centres and stores in the area, including retail floorspace, again distinguishing between convenience and comparison shopping. Appropriate turnover/floorspace ratios are then applied to estimate existing turnover, based on observed trading performance and typical company averages.

Step 4: turnover of proposed development

The turnover of the proposal is normally estimated on the basis of company averages if there is a known retailer, or the trading performance of similar stores elsewhere and the trading potential of the catchment area.

Step 5: assessment of trade draw

It is necessary to estimate the source of the turnover of the new store in terms of its percentage draw from residents in different parts of the catchment area, usually isochrones. Inflows must also be considered; a proportion of turnover will come from outside the primary catchment area. Some consultants use the concept of 'market penetration' to estimate the likely share of total trade that a particular type of store could achieve based on experience elsewhere.

Step 6: estimate of trade diversion

Knowing from which areas the trade is likely to be drawn, the amount of trade diverted to the new store can then be represented as a trade loss from existing stores, taking into account any clawback of leakage. In most defined market areas there is some leakage of trade to external centres/stores. Trade diversion is expressed as a percentage of the turnover of existing centres in the design year, and some consultants also assess the residual turnover of centres after trade loss has taken place. Cumulative impact may also have to be considered.

Step 7: implications of trade diversion

It is important to assess the ability of affected centres to withstand impact in terms of the effect on their vitality and viability, and the effects on future investment. This is a qualitative rather than quantitative element of the assessment, but it is extremely important in making a professional judgement about retail impact. The final planning decision on a proposed development will often be based on this qualitative judgement.

Norris (1992) who has carried out detailed research into the application of RIA in relation to regional shopping centres, concludes that the 'simple and transparent step-by-step manual approach still provides the most appropriate framework for improving on the existing impact methodologies'. Using this framework, he put forward 'a common-sense and pragmatic approach to the impact problem . . . which encourages the use of reasonable and robust assumptions, and attempts to reduce the level of professional "bias" involved at each stage of the analysis'. Although this approach was developed for application to regional shopping centres, it is generally applicable to all types of retail development.

Norris put forward a 'refined quantitative approach' to the assessment of 'need' and 'impact' in which he set out the main steps in the methodology and a number of suggested changes to bring about improvements, as shown in Table 4.1.

Table 4.1 The refined quantitative approach

Main steps in the methodology	Suggested changes to the methodology
Delineation of catchment area	No change
Population estimates	No change
Expenditure per capita estimates	No change
Expenditure growth forecast	Test a range of assumptions
Base year turnover estimate	No change
New commitments	A need for greater agreement at pre-inquiry meetings
Floorspace 'efficiency' growth	Test a range of assumptions
Design year turnover estimate	A need for greater agreement at pre-inquiry meetings
'Need' assessment	A need to give greater weight to qualitative assessment
Determination of centre/store turnover Determination of centre trade draw	A triangulation method
Trade diversion	An interactive method
Impact forecast	A need to adopt a consistent approach which compares like with like

The two key elements of the assessment of impact inherent in Norris' approach are the determination of turnover and trade draw, and the determination of trade diversion.

Norris suggested a 'triangulation method' in which the sensitivity of the turnover and trade draw estimates can be tested using three different control mechanisms:

- Step 1 – use the conventional 'disaggregate' approach based on company average turnovers
- Step 2 – estimate the market share which the store could be expected to take
- Step 3 – analyse the socio-economic characteristics of the catchment area.

In determining trade diversion, Norris recommended detailed household surveys to build up an expenditure flow matrix, and suggested an 'interactive method' which 'combines the transparency and simplicity of the manual model with the greater flexibility and rigour of the gravity model'.

Like other recent advice on impact assessment methods, Norris advocated an approach to the assessment of regional centre proposals which synthesises quantitative and qualitative methods. The approach comprises a broad assessment of the vitality and viability of centres that may be affected, then a more detailed assessment of the most vulnerable centres following the quantitative assessment of need and impact.

Evaluation of approaches

Data and assumptions

Catchment areas

The concept of a 'catchment area' is quite different from an arbitrary 'study area'. It implies the definition of an area within which shopping patterns are known, based on information about shoppers' behaviour. There are two ways of determining a catchment area:

First, using household survey data to establish where people shop, usually distinguishing between main food shopping, top-up food shopping, and non-food or comparison shopping. It is conventional to define the primary catchment area from which most trade (usually about 80 per cent) will be drawn and the secondary catchment area which is wider and includes more distant trips.

Secondly, defining isochrones at five- or ten-minute intervals from the location of the proposed development. Isochrones are useful because there is evidence from post hoc studies to show the typical percentages of trade drawn from different time bands for different sizes and types of store.

In practice, it is quite common for a primary catchment area to be defined initially on the basis of isochrones but then modified to take account of the overlapping catchments of competing centres. In other words, the definition of the catchment for a proposed development should be as realistic as possible in the particular local circumstances. Although the determination of a primary catchment area is essential to provide the spatial system for quantitative analysis, the precise demarcation of boundaries is not critical as long as data for the demand side (expenditure) and supply side (turnover) of the analysis are compiled for the same geographical area.

In simple applications it is possible to regard the catchment area of a proposal as one spatial unit, though it is usual to define isochrones in order to assess trade draw from different time bands. For more complex proposals in urban areas, one can adopt a more detailed approach, subdividing the catchment area into sub-areas which are based on isochrones disaggregated into smaller zones to give a more accurate representation of shopping trip patterns and facilitate the distribution of trade draw.

Expenditure

Expenditure in the catchment area (and sub-areas) is calculated by multiplying the population by a figure of expenditure per head. Population for the base year is generally available from census-based figures for small areas, and population projections for the design year can be obtained from local authorities, although there can sometimes be problems in estimating future populations for areas below the level of local authority districts. It is best to use local authority forecasts for small areas (wards or parishes) if possible because they reflect the actual distribution of new housing in the area, rather than assuming that it is spread evenly.

Estimates and projections of per capita expenditure are much more problematical, and this is often an area of debate and disagreement between parties in assessing the impact of a proposed development. The starting point is usually to take a base figure of per capita expenditure. Expenditure data has been available for many years from the Unit for Retail Planning Information (URPI). (From 1997 to 1999 URPI was known as The Data Consultancy, and since 1999 it has operated as MapInfo Ltd.) In this book both the names URPI and The Data Consultancy are used depending on the date of publication of the relevant data. Per capita expenditure figures are available from The Data Consultancy either as national average figures updated annually in the information briefs or (preferably) by commissioning an Illumine demographic report for a defined local area such as a local authority district or aggregations of wards. The Illumine local expenditure data has the advantage that it takes account of the local socio-economic structure of the catchment area, which is important in areas which are likely to deviate from typical national average socio-economic conditions, e.g. areas of high unemployment.

Handling per capita expenditure data involves two major problems: the choice between goods-based and business-based figures, and the choice of the most appropriate growth rate to apply to project expenditure in the design year. The definition of goods-based and business-based expenditure is shown in Table 4.2.

It was conventional until quite recently to use expenditure figures defined by goods type. Therefore, the URPI/Data Consultancy figures for per capita expenditure on convenience or comparison goods would be adopted as the most reliable basis for calculating the demand side of the impact assessment. Most consultants, however, are now in favour of using expenditure by business type. Business-based figures, which are also available from The Data Consultancy, have been adjusted so that they represent spending in convenience or comparison businesses, rather than on particular goods categories. The differences between goods-based and business-based expenditure can have significant ramifications for the assessment of impact. For instance, a superstore is a convenience business but it will sell some comparison as well as convenience goods. If goods-based convenience expenditure figures are being used, the turnover of the store must be adjusted to exclude the comparison goods element of the total turnover, otherwise the demand and supply side definitions will be inconsistent.

Table 4.2 Expenditure by convenience and comparison goods and business categories

Goods based	Business based
Convenience goods	*Convenience businesses*
Food (household expenditure)	Stores selling food, beverages and tobacco
Alcoholic drink (proportion spent in retail outlets)	
Tobacco (proportion spent in retail outlets)	
Other goods (newspapers and magazines; cleaning materials)	
Comparison goods	*Comparison businesses*
Clothing and footwear	Stores specialising in the retail sale of other new goods
Do-it-yourself goods	
Household goods (furniture, pictures, etc.; carpets and other floor coverings; major appliances; textiles and soft furnishings; hardware)	Personal and household goods repairers
Recreational goods (radio, television and other durable goods; television and video hire, but excluding licences and repairs; sports goods, toys, games and camping equipment; other recreational goods; books; bicycles)	Stores selling second-hand goods
Other goods (pharmaceutical products and medical equipment; toilet articles and perfumery, jewellery, silverware, watches and clocks; other goods)	Stores specialising in pharmaceutical and medical goods, cosmetic and toilet articles
	Other non-specialised stores

Source: The Data Consultancy.

Use of expenditure by business type has the advantage that it simplifies the impact analysis by removing the need to adjust the figures to exclude comparison goods, floorspace and turnover. But the use of business-based estimates raises the additional problem mentioned above – it affects the growth rate that should be applied to forecast future expenditure. Business-based estimates imply higher growth rates than those for goods-based expenditure because of increased diversification into non-food goods by food retailers, particularly the big four operators – Sainsbury's, Tesco, Asda and Safeway. These companies have thus been able to benefit from the more

Table 4.3 Long-term growth in per capita expenditure, 1976–1998 (per cent per annum)

	Goods-based	Business-based
Convenience	0.1%	1.8%
Comparison	4.2%	3.6%

Source: The Data Consultancy (1999a and 1999b).

rapid growth of comparison goods expenditure, and much of their recent growth is due to sales of non-food goods. Non-food items accounted for 25 per cent of all grocery outlet turnover in 1996. There is likely to be an increasing trend towards the growth of comparison goods sales within large food stores affecting all major operators.

Long-term growth rates for goods-based and business-based per capita expenditure are shown in Table 4.3. The trends in expenditure growth for convenience and comparison goods are shown in Figure 4.3.

There is not a great deal of difference between the figures for comparison goods and comparison businesses. Both show relatively high growth rates.

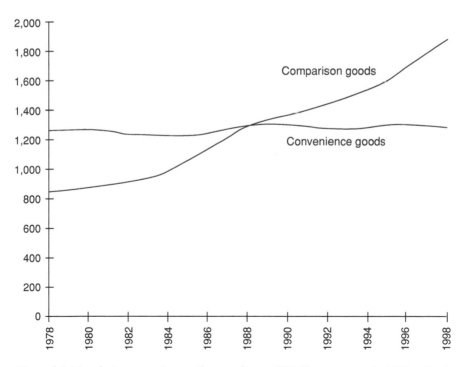

Figure 4.3 Trends in per capita retail expenditure, UK (£ per annum in 1995 prices)

But for convenience goods and businesses the differences in the growth are quite significant. The application of the convenience goods-based rate over a five-year forecasting period would produce a 0.5 per cent overall growth, but the application of the business-based rate would result in a 9.3 per cent compound growth. The choice of goods or business base, then, could have an important bearing on the capacity of the retail market to accommodate additional development without adverse impact on existing shops. The choice of the higher-growth business-based forecasts by some consultants acting for clients seeking to promote new development is perhaps not surprising. For this reason it may be advisable to make expenditure forecasts, particularly when dealing with a superstore proposal, using both goods-based and business-based figures, treating them as a range of forecasts.

Another decision to be made in projecting expenditure is to select the most suitable trend in past expenditure growth to use as the basis for the future projection. For both goods-based and business-based projections, The Data Consultancy provides data on short-term, medium-term, long-term and ultra long-term trends. For goods-based expenditure the latest advice is that the long-term growth rates (1976–1998) or rates between the long-term and the ultra long-term (1963–1998) trends are the most realistic (The Data Consultancy, 1999b). However, there are problems in using the goods-based figures for convenience expenditure. Convenience goods retail expenditure per head has shown a very irregular pattern over the past 40 years, with very little real growth, producing low correlation coefficients and making projections subject to considerable unreliability. The best fit to the data is the ultra long-term trend of 0.1 per cent per annum but even this is not statistically reliable (The Data Consultancy, 1999b).

For business-based expenditure the long-term trends also have the highest correlation for comparison store businesses, but 'for convenience store businesses the short-term growth rate now suggests that a slightly higher growth rate than the long-term growth rate might be sustainable' (The Data Consultancy, 1999a: 4).

A further complication in any expenditure analysis is the need to adjust the figures to exclude 'special forms of trading' such as mail order and automatic vending machines. This represents about 1 per cent of expenditure on convenience goods and 8 per cent on comparison goods. Best practice advice on expenditure projections is given in Chapter 5.

Turnover of existing centres

Estimating the turnover of existing centres has been difficult since the demise of the Census of Distribution, last held in 1971. Sparks (1996: 93) has described the demise of the Census of Distribution in Great Britain as a disaster for local authority planners, retailers, developers and academics alike. He said 'the hypocrisy of a government issuing planning guidance suggesting the key questions are vitality, viability and impact, but then denying the need to collect data to answer such questions is breathtaking'.

During the 1970s and even into the 1980s it was common for consultants to estimate turnover by trying to update the 1971 Census of Distribution figures, but it is no longer acceptable to do so. A number of suggestions have been made for estimating retail turnover. For instance, Guy (1984) attempted to estimate the retail turnover of shopping centres in Cardiff by examining the relationships between convenience trade (based on survey data), number of visits to a store, and the store's sales area.

The most common method of estimating turnover, however, is to apply ratios of turnover/floorspace to the observed floorspace of individual stores or to the aggregate floorspace of different categories of shop. The method can be quite successful if accurate floorspace data are available and if one has confidence in the turnover/floorspace ratios to be applied. For larger stores, such as superstores and multiple comparison goods retailers, annual company reports give details of total turnover and floorspace. There are several sources of data on average turnover ratios of leading retailers, such as *Retail Rankings* produced by Corporate Intelligence on Retailing. The main requirements in estimating the turnover of a centre are:

- to obtain a reliable measurement of net floorspace for the major retailers and for smaller shops by type of shop
- to make a judgement about whether the centre in question is trading above or below average performance, and decide on appropriate figures of turnover per square metre (or square foot)
- to apply the turnover/floorspace ratios in as disaggregated a way as possible.

For foodstores, the Institute of Grocery Distribution Stores Directory provides floorspace data supplied directly by grocery retailers but does not include discount food operators or independent retailers. This is a useful source for the larger foodstores but in general it is best to use local authority floorspace survey data where it is available – because of its comprehensiveness and because it is more likely to be readily acceptable to the local authority than other sources of floorspace data.

Using available floorspace data and reliable estimates of turnover per square metre, a reasonably accurate picture should be built up of existing turnover. The exercise should be carried out as part of the essential task of assessing the existing performance of all shopping centres in the catchment area. It requires a qualitative as well as quantitative approach to assess the vitality and viability of each centre. Where household survey information is available on shopping patterns it may be possible to estimate the turnover of existing centres and large stores on the basis of their market shares of retail expenditure in the survey area. This method can be very reliable if used with care, and its practical application is discussed in Chapter 5 in recommending best practice in RIA.

Turnover of a new retail proposal

'The turnover of any proposed store is the single most important variable in assessing its impact' (Roberts, 1982: 9). Turnover is the product of two variables: the number of transactions and the average sum spent by each customer. The simplest approach, according to Roberts (1982), is to try to relate turnover to the population within the catchment of the store. The likely turnover is very much dependent upon the level and distribution of the population it has been located to serve. Drivers Jonas (1992) also advises that the turnover of a proposed new store should be made with reference to the characteristics of the catchment area, e.g. population density and available expenditure, as well as by consulting published statistics on the performance of various retailers.

The use of company averages is an important reference point, but the figures are averages and there is very considerable variation in the performance of individual stores within the same chain. The use of figures for similar stores can also lead to problems because there are inconsistencies in the published figures, e.g. in the treatment of VAT and petrol sales at superstores. Sometimes, for the purposes of a planning inquiry, a developer or major retailer will put forward its own 'accurate' estimate of a new store's turnover. There is a natural tendency for a developer or retailer to argue for a lower level of potential turnover than the local planning authority. But, even if the developer and local planning authority are able to agree on potential turnover, Howard (1986) notes that evidence given in public inquiries on the level of turnover expected to be generated has not proved to be a reliable guide to subsequent actual turnover. A good example of this is in Noel's case study (1989) of the proposed Tesco superstore development at Neasden, which opened in 1985. He shows that the consultants for Brent Council and for Tesco both underestimated the store's turnover compared with surveys carried out by the Greater London Council after the store opened.

Trade draw

'Trade draw' is the proportion of the estimated turnover of a proposed store which is derived from the catchment area and subdivisions of it. The trade drawn is normally expressed as a percentage of the estimated generated spending from each isochrone and from outside the primary catchment area. The concept of 'market penetration' is sometimes used to represent the percentage of a store's turnover drawn from a particular isochrone. The concept of 'market share' is also commonly used to represent a store's turnover as a percentage of spending in the primary catchment as a whole. The allocation of trade draw between isochrones is one of the most difficult steps in the assessment and it requires assumptions to be made based on professional experience.

In assessing trade draw it is necessary to look at retention levels and leakage. The retention level of an existing centre is calculated as the turnover of that centre expressed as a proportion of the available retail expenditure in the catchment area. This gives a net figure, ignoring flows of expenditure out of or into the catchment area. A retention level of more than 100 per cent indicates a net inflow of expenditure and can be taken as evidence of a centre which is drawing trade from a wide area, typically a sub-regional centre. A retention level of less than 100 per cent indicates a net outflow and is typical of a small town centre or a district centre in a catchment area where there is leakage to a larger town or city centre nearby.

Leakage of spending from a primary catchment area may be an important issue in looking at the impact of a proposed new development. A new superstore, for instance, may have the effect of considerably reducing leakage out of the area. It is usual in carrying out impact assessments to assume that a new development will result in clawback of some of the spending that is currently lost as leakage. This can also be a sensitive issue in assessing impact because trade that is derived from clawback has the effect of reducing the amount of trade that is diverted from existing centres within the catchment area. It is, therefore, in the interests of the promoter of a new development to argue that it will result in a significant amount of clawback.

Trade diversion

Trade diversion from existing centres is the crucial and most contentious element of impact assessment. It is contentious because it involves a considerable amount of judgement (Roberts, 1982). 'The 'science' of retail impact analysis rapidly transforms into subjective assessment at this stage in the process (Drivers Jonas, 1992: 77).

Percentage trade diversion is usually based on subjective judgements taking account of existing shopping patterns. One approach is to assume that each store or centre will retain its relative share of trade after the loss of trade to a new development, but this is too simplistic. It is more reliable to develop a matrix of flows of expenditure between sub-areas (within isochrones) and centres in the base year and project this matrix forward to the design year allowing for increases in expenditure and turnover. The estimated turnovers for each centre are used as the denominators in the subsequent calculation of percentage impacts. The proposal is then superimposed on the design year expenditure flow matrix and the estimated trade draw of the proposed store in terms of expenditure from each sub-area is then subtracted from the available expenditure. The remaining expenditure is allocated between existing stores and centres, and the difference between the 'pre-impact' and 'post-impact' turnovers represents trade diversion from each centre, expressed in percentage terms. The practical application of this matrix approach is covered in detail in Chapter 5.

This matrix method also enables cumulative impact to be assessed. Each proposal can be assessed individually, and these can be combined to show their cumulative effects. It is conventional in such cases to allow for a reduced turnover for the new stores because they will compete with each as well as with existing centres. In addition to percentage trade diversion, it is usually important to consider the residual turnover of shopping centres and make a judgement about whether the level of residual turnover is adequate to maintain the viability of a centre.

Critical analysis

Deficiencies in approaches

A frequent criticism of RIA in the past has been the lack of consistency in the methods used at planning inquiries. There are still deficiencies in the methods used and the key deficiencies were pointed out by Drivers Jonas in its review of the methodologies, as follows:

- lack of analysis of current shopping patterns and the strengths and weaknesses of existing centres
- lack of explanation of assumptions in apportioning trade diversion
- lack of attempt to establish a relationship between available expenditure and potential turnover
- limited interpretation of impact estimates
- insufficient attention given to residual turnover and the level of reliance of a shopping centre on a particular type of shopping trip
- assumptions about likely store closures are often unjustified
- lack of detailed examination of the functional relationship between a new retail development and existing centres
- disputes about data which can undermine confidence in an RIA and often can be avoided through negotiation (Drivers Jonas, 1992).

Retail impact assessments should be accurate with limited sensitivity, thorough, consistent and capable of agreement between parties. Local planning authorities should make available to consultants all data held by the authority which is likely to be of assistance to the consultant in undertaking a RIA (Drivers Jonas, 1992: ix).

Data availability

The Drivers Jonas review notes that the availability of population and expenditure information does not generally give rise to problems, but there is a lack of comprehensive information relating to retail floorspace, turnover and local shopping patterns. These problems have arisen largely because of the cancellation of the 1981 Census of Distribution. There is now no locally based information on shops or shopping centres, including turnover, floorspace and retail employment. The main data deficiencies according to Wade (1983) are local estimates of retail turnover and local estimates of floorspace.

The Census of Distribution was replaced by an annual Retail Inquiry which started in 1976 based on a sample of retail businesses. But from 1981 it has become a biennial full inquiry with a much smaller survey collected in the intervening years. It does not provide the range of information previously available from the Census of Distribution. Most crucially, it does not provide area statistics owing to the sample nature of the inquiry, and to the business not retail outlets being the reporting unit (Dawson and Sparks, 1986).

Shopping surveys

The lack of local data, at least on shopping patterns, can be overcome by carrying out surveys. There are two kinds of surveys of shopping patterns: shopping centre or street surveys, and household surveys. There are many examples of the use of both types of survey. For instance, Robson (1987) carried out a local survey of shopping habits and opinions in eight centres in north London. Data derived in each of the centres were related back to home address, enabling catchment areas to be drawn for each centre. Sherman and Dossett (1995) carried out a county-wide survey of retail capacity, shopping patterns and customer desires in Cheshire in 1994. The survey showed how it is possible to gain significantly more information about shopping behaviour by surveying at the sub-regional, rather than local, level. Parsons and Sherman (1994b) recommend ongoing monitoring of activity patterns through regularly updated surveys to provide an understanding of how the retail system works as the best rational basis for policy. Valuable advice on designing and undertaking store and street interview surveys is provided by URPI in Information Briefs 90/8 and 91/2. Guidance is given on obtaining a balanced sample of responses. The use of household surveys in particular is highlighted in Chapter 5.

Accuracy

One of the key requirements of the quantitative assessment of impact is accuracy.

> This requirement relates not merely to the correct use and application of data such as expenditure per head and population, but also the assumptions made regarding shoppers' behaviour. There are

strong reasons for the use of surveys to investigate existing shopping patterns as a basis for predicting future change.

(Drivers Jonas, 1992: 119)

Data problems and assumptions can raise doubts about the accuracy of
the results of RIA. The results are very sensitive to statistical uncertainty.
McCallum (1995) points out that RIAs should not lay claim to an accuracy
that cannot be justified and more weight should be given to the assessment
of existing town centres. Guy also makes useful comments about the accuracy of methods:

> The accuracy of any forecasting method is limited by the degree of
> truth in its assumptions about human behaviour and by the quality
> of the data used. The worse the assumptions or data, the less merit
> there is in precise 'calibration' or in overcomplexity of method. The
> most useful type of method is one where the results are simple in
> character, relatively insensitive to minor variations in assumptions
> or data, and capable of being produced quickly once the basic data
> have been assembled. The main advantage of this type of method is
> not in the 'accuracy' of its forecasts but in its ability to be used in
> comparing alternative proposals from a consistent standpoint.
>
> (Guy, 1977: 501)

Impartiality

In addition to criticisms of the accuracy and reliability of impact studies,
Noel (1989) comments that they usually fail the test of objectivity because
they are prepared by a planning authority who have to defend a refusal of
planning permission at appeal, or by a consultant representing a developer/
retailer who will seek to argue that the proposal will not have a significantly
damaging impact. 'Accordingly there is considerable room for assessing the
accuracy and reliability of impact appraisals in an impartial manner' (Noel,
1989: 14).

It is important that a RIA should have the confidence of those who will
rely on its conclusions in order to make planning decisions. The Drivers
Jonas review states that the greater the level of agreement between the
parties as to the input data which are to be used, the less will be the scope
for manipulation of results.

> The most common areas of dispute are on the levels of trade draw
> from different sectors or drive-time bands, turnover of the proposed
> development, turnover of existing centres, and the relative propor-
> tions of trade likely to be diverted from each centre. If disagreement
> can be narrowed down to the very last stage of trade diversion

estimates . . . the dispute then focuses primarily upon the subjective judgements which inevitably have to be made regarding the relative attractiveness of centres, their exposure to competition, and the likely changes in shoppers' behaviour.

(Drivers Jonas, 1992: 110)

Cumulative impact

Government advice in PPG6 refers to the need to consider the cumulative impact of recent and proposed retail developments. Often it is necessary to take account of committed proposals, that is stores which have been permitted but not yet built, or at appeals to assess the impact of several proposed stores. The most common approach is to total the combined estimated turnover of the new proposals and then apply a discount to that total because competition between them will reduce each one's turnover potential. If two stores are being developed, their impact on each other could reduce the combined turnover to no more than 85 per cent of the stand-alone turnover; if three stores are being developed, turnover levels may be no more than about 75 per cent of the stand-alone turnover.

Sectoral analysis

RIAs tend to focus on particular sectors of the retail market rather than on retail trade as a whole. In assessing the impact of a supermarket, for instance, the analysis will usually be carried out in terms of convenience trade only. The percentage trade diversion will be in terms of convenience trade not the total trade of the centre. This approach can give a misleading impression. The overall impact on a centre, especially one with an important comparison shopping role, will be much less than the impact on convenience trade only. In practice, although it is necessary to assess impact separately for convenience and comparison shopping, in some situations it is advisable also to consider the overall impact in terms of total trade. Many RIAs are carried out for foodstore proposals – supermarkets and superstores. The question arises whether the methods that have been developed and used for foodstores can be applied to non-food retail developments. The differences in approaches for different types of retail development are discussed in Chapter 7.

Interpretation of impact

The figures produced by quantitative impact assessments in themselves mean very little. It is the effects on actual patterns of trade that are important in planning terms (Noel, 1990). Professional judgement must be applied, particularly in dealing with impact issues at inquiries. Inspectors will attach more weight to the interpretation of the figures than to the figures themselves.

Drivers Jonas lists a number of key factors which should be examined when interpreting the results of a RIA, in order to establish whether there is a serious risk of failure of an existing shopping centre or a threat to its function. The key factors are the existing conditions and the effects of the new development. The *existing conditions* include the role of the centre, how it is performing, and whether it is improving or declining. The *effects of the new development* include the implications of the levels of residual turnover, the threats to any vital 'anchor' retailers, the prospects for reletting if there are closures or vacancies, the effects on committed developments or future investment, the trading impact on the centre as a whole, i.e. total trade, the role of other (non-retail) uses in the vitality of the centre, the location of the proposed development in relation to the centre, the provision of new community facilities in the new development, the prospects of improvements in other shops to remain competitive, and the prospects of an overall improvement in the range and quality of shopping facilities (Drivers Jonas, 1992).

Interpreting the significance of a percentage impact on a particular centre is very difficult. For a long time local authorities tended to apply a '10 per cent rule' that, if the impact of a proposed development was more than a 10 per cent trade diversion, it was considered unacceptable. It is generally true to say that single-figure impacts are usually regarded as acceptable by local authorities and inspectors, though there are some notable exceptions where appeals have been dismissed in situations where the figure was below 10 per cent but there were concerns about the vitality and viability of the centre in question. Double-figure trade diversions are a cause for greater concern, especially if the figures are over 20 per cent. In such cases consultants may claim that the centre is over-trading or is sufficiently strong to withstand such an impact. Hence the importance of considering a wide range of quantitative and qualitative factors in interpreting impact. Residual turnover is usually given less attention than percentage trade diversion in making planning decisions but it is an important indicator of the post-impact viability of a centre (Drivers Jonas, 1992).

The Drivers Jonas review also states that there appears to be very little evidence to show a causal relationship between the opening of new retail developments and the closure of existing shops except where existing retailers relocate into a new development. There may be time-lags before the effects of trading impact are manifested in closures or vacancies. Independent traders may continue to operate even though their businesses are not viable, or retailers may operate under long leases that discourage them from closing unprofitable branches in town centres until the end of their lease period. This is a point recognised in PPG6, hence the advice that in assessing impact it is essential that the local authority takes a long-term view.

The revised PPG6 also, for the first time, specifies the factors to be considered in assessing applications for retail development which may have an impact on a nearby town, district or local centre. These are as follows:

the extent to which developments would put at risk the strategy for the town centre, taking account of progress being made on its implementation, in particular through public investment

the likely effect on future private sector investment needed to safe-guard the vitality and viability of that centre

changes to the quality, attractiveness and character of the centre, and to its role in the economic and social life of the community

changes to the physical condition of the centre

changes to the range of services that the centre will continue to provide, and

likely increases in the number of vacant properties in the primary retail area.

(DoE, 1996, para. 4.3)

This list is similar to the key factors listed above which were identified by Drivers Jonas, and it is helpful to those involved in assessing retail impact as it clarifies how the quantitative analysis of impact should be interpreted using qualitative factors. PPG6 notes that the information collected on the health of town centres using indicators of vitality and viability should help in undertaking such an assessment.

Summary

There is a general lack of understanding in the planning field about the methodology of RIA and how it should be applied in practice. In dealing with proposals for new retail development, decision-makers are concerned with quantitative and qualitative need, and the location of the development, as well as its impact on existing centres.

Three main approaches have been developed over the years to assess retail impact: post hoc studies, shopping models and predictive impact assessments. Post hoc studies of the trading effects of retail developments using questionnaire surveys were common in the 1970s on early superstore developments. Shopping models were widely used in Britain in the 1960s and early 1970s but they were beset by theoretical and operational problems, particularly concerned with calibration. Further developments in shopping models took place in the 1970s, notably by URPI.

As the use of models has diminished since the 1970s, step-by-step approaches have been increasingly used to carry out predictive impact assessments. In the last decade the step-by-step method has been refined. A fairly common methodology is now used by consultants, but with some variations. There are six main steps:

- definition of the catchment area of the proposed development and sub-areas based on isochrones and zones
- estimation of expenditure based on population and per capita expenditure for the base year and projections for the design year – either by goods type or business type
- estimation of the turnover of existing centres by applying turnover/floorspace ratios to retail floorspace data
- estimation of the turnover of the proposed development
- assessment of trade draw from within the catchment area and beyond
- estimation of percentage trade diversion and (sometimes) residual turnover.

An evaluation has been made of the data and assumptions involved. It is particularly important to:

- make careful use of expenditure data and projections
- estimate the turnover of the proposed development accurately
- use appropriate assumptions on trade draw
- consider residual turnover as well as percentage trade diversion, and
- assess the ability of centres to withstand trade loss.

A critical analysis of the methodology indicates:

- deficiencies in the approaches
- problems of data availability
- the value of shopping survey data
- the requirement for accuracy and impartiality in the assessment
- the need to consider cumulative impact
- the relevance of assessing overall impact as well as sectoral analysis.

In the past the emphasis of approaches to RIA was on the quantitative assessment of impact. However, it is no longer appropriate simply to assess impact in quantitative terms; there also needs to be a qualitative assessment of the implications of trade diversion. The results of any quantitative impact assessment must be subject to careful interpretation, using professional judgement. The key factors to be considered are the existing conditions of the shopping centre and the effects of the new development itself. Interpreting the significance of a percentage impact on a particular centre is very difficult. Government guidance in the form of PPG6 now explicitly recognises that the impact of a proposed development must be judged against the health of the shopping centres that are likely to be affected, but there is still no clear guidance on how retail impact should be assessed in practice.

5

A FRAMEWORK FOR RETAIL
IMPACT ASSESSMENT

In this chapter some principles for best practice in RIA are laid down. They are rooted in the policy requirements of PPG6 and the technical requirements for greater objectivity and a more informed use of data and assumptions in quantitative assessment. A general framework for the application of RIA is presented which complements the policy guidance in PPG6. This chapter recommends best practice in the light of the current policy context for retail planning. Recommendations are made for improvements in quantitative assessment using an expenditure flow matrix approach, which has been tested extensively in practice, to analyse existing shopping patterns and to predict changes in shopping patterns resulting from proposed new retail developments. Qualitative assessment is discussed in Chapter 6.

Improving the application of retail impact assessment

The need for advice on best practice

There are still inherent problems with the application of RIA in Britain. There is clearly a need for improvement in both quantitative and qualitative assessment. Approaches to RIA have not kept pace with changes in the policy context in the last decade. It is still a predominantly quantitative process which concentrates on economic impact rather than other relevant factors such as social impact and the growing concern with sustainability and environmental factors.

The most important research on retail impact in Britain in recent years has been:

- research commissioned by the Scottish Office on RIA methodologies, which was carried out in 1992 by Drivers Jonas
- the PhD research by Steven Norris at Reading University on the use of impact assessment in studies of proposals for new regional shopping centres, also completed in 1992

- the PhD research by the author at Newcastle University, which was a critical examination of the application of RIA in the planning process (England, 1997).

The research by Drivers Jonas is covered in detail in Chapter 2. Drivers Jonas examined the effectiveness of common approaches to RIA and identified key elements which might form the basis for an optimum method of assessment. The research was of particular reference to Scotland but it is of more general relevance to the application of RIA in Britain. It found that there are many similarities in approaches to RIA but there are variations in the approach to the later stages of the process. The report drew attention to the fact that percentage impact estimates are of limited use and have to be examined 'on the ground'. A good understanding of existing shopping patterns is also important.

Norris' research is referred to in Chapter 4. Norris (1992) concluded that the main approaches to RIA have changed very little since the mid-1960s and are still subject to a 'myriad of indeterminate assumptions and considerable areas of doubt'. These uncertainties have been compounded by the poor information base on which the methods are based. There is still a preoccupation with developing more sophisticated techniques and achieving greater statistical accuracy, often at the expense of the important qualitative stage of the analysis. Norris felt that current practice needs to be improved in three fundamental areas:

- the need for a stronger and more comprehensive retail information base
- the need for a better trained, better informed and more critical Planning Inspectorate
- an improvement in impact methodologies and a more pragmatic approach, including an improvement in qualitative analysis.

Drivers Jonas and Norris based their conclusions largely on experience of the 1980s but what has happened in the subsequent years to influence the application of RIA? The author's PhD thesis drew attention to other and more recent clear evidence that the approach to RIA needs to be improved. The main sources of evidence, which are covered briefly in this section, are:

- comments by local authorities on retail impact studies
- the report of the House of Commons Select Committee on the Environment
- the views of other researchers and consultants
- the attitude of planning inspectors.

Comments by local authorities

The survey of local authorities, which is covered in detail in Chapter 8 of this book, revealed some significant criticisms of RIAs, in particular:

- the need for better/more up-to-date retail statistics at a local level
- the need for RIAs to be more independent/objective/impartial
- the need for surveys of local shopping patterns/catchment areas
- less bias in favour of the proposed development
- more co-operation between the applicant and the local authority.

Other criticisms mentioned by some local authorities include more emphasis on vitality and viability indicators, the need for a standard/more consistent methodology, and a range of scenarios/assumptions (sensitivity). From the local authority viewpoint, therefore, the overall quality of RIAs needs to be improved.

The Environment Committee

The House of Commons Select Committee on the Environment report, 'Shopping centres and their future', in October 1994 called for clearer and more detailed retail planning guidance for planning inspectors and local authorities, especially on the anticipated impacts of proposed retail developments and on long-term social and environmental effects. The committee recommended that such guidance be brought together in a 'handbook'. In its response to the Environment Committee's report the government agreed that it is important that planning policy guidance should be contained in a single document, but did not accept that it is necessary or desirable to create a handbook. Therefore the opportunity to clarify how retail impact issues should be handled by local authorities was missed, and the need for such advice on best practice remains. The Environment Committee issued a supplementary report in March 1997 which called on the government to insist that full impact studies should accompany all applications for significant retail development, particularly in or around small or market towns (para. 20), and recommended that in cases where development proceeds, impact studies should be assessed so that their accuracy can be reviewed and any necessary action taken to improve future studies (para. 21) (House of Commons, 1997).

Other views

Several researchers and consultants working in the retail planning field have commented on the need for improvement in the application of RIA. For instance, McCallum (1995) recognised that there is a clear need for some measure of likely impact to make informed decisions. He thought the development of best practice guidance might help and would save Public Inquiry time. He said that RIAs should not lay claim to an accuracy or reliability that cannot be justified and more weight should be given to the assessment of existing centres. A report by Healey and Baker (1995) highlighted the need for further research on approaches to RIA and their validity. It stated that, if possible, a preferred approach should be identified or a good practice guide provided to ensure that such assessments are carried out on a consistent and readily understood basis. Norris and Jones stated that:

> Impact assessment can never be a scientific activity, as there is no watertight method or comprehensive retail planning data to support such a claim. Thus we firmly believe that the excessive technical and pseudo-scientific language which dominates current practice needs to be replaced by a more commonsense and pragmatic approach.
>
> (Norris and Jones, 1993: 85)

Retailers have also expressed concern about the quality of RIAs. The CB Hillier Parker research study for the DETR on the impact of large food stores on market towns and district centres shows that there is a general consensus among retailers such as Tesco, Sainsbury's and Somerfield that a good practice guide is needed to ensure the consistency and comparability of RIAs.

> There is a pressing need for a common basis for assessment. We consider that this would assist greatly in enabling local authorities and the private sector to work together more effectively, and reduce unnecessary time and cost at public inquiries spent deliberating issues which could and should be dealt with at an earlier stage.
>
> (CB Hillier Parker, 1998: 93)

The author's thesis highlighted the scepticism with which RIA is held by planning inspectors. There has over many years been an explicit or implicit criticism by inspectors of RIA, particularly of quantitative approaches. This is a matter of some concern because a significant number of decisions on proposed developments are made on appeal and impact is still an important factor in appeal decisions. The Environment Committee thought that planning inspectors should achieve greater consistency in decision-making. The latest government guidance in PPG6 gives only very general advice on retail impact issues, and there is still no clear advice to inspectors in this field.

Policy requirements

PPG6 raises some crucial points about shopping policy and the assessment of proposed retail developments. RIA takes place within a policy context and the policy requirements underlying impact assessment have never been so important as a basis for planning decisions. In this section the guidance in the latest version of PPG6 is analysed to draw out the implications for best practice in the application of RIA. The main guidance relating to RIA is summarised in Table 5.1 which refers to the relevant paragraphs in PPG6 (England, 1996).

The revised guidance goes some way towards clarifying how the impact of new retail developments should be assessed. It does not make any recommendations on the methodology to be used to assess economic impact but it provides useful advice on the factors to be considered in relation to proposals for retail development.

The key tests in PPG6, shown in Table 5.1, are intended to apply to all new retail developments that are proposed outside existing centres. In the case of proposals for retail development over 2,500 square metres gross floorspace, applications must be supported by evidence of impact. Impact assessments may also be necessary for smaller developments depending on their size and nature in relation to the centre, e.g. in a market town. Where such evidence is required, PPG6 spells out the information that should be provided.

PPG6 has other policy requirements which are additional to these three key tests. The first requirement for a developer proposing an out-of-centre development is to show that he has followed the sequential approach laid down in PPG6. 'The onus will be on the developer to demonstrate that he has thoroughly assessed all potential town centre options' (DoE, 1996, para. 1.9). The sequential approach also implies that the local authorities should use RIA at the local plan formulation stage before sites are allocated. As part of the sequential approach the need for a development must be demonstrated by means of a capacity analysis. 'If . . . there is no need or capacity for further developments, there will be no need to identify additional sites in the town' (para. 1.10). The local authority will then need to be satisfied about impact on the development plan strategy. A proposal 'should be assessed against the strategy in the development plan and be refused if it would undermine that strategy' (para. 4.2).

Reference is also made to environmental impact. Major shopping proposals, where appropriate, may have to be supported by evidence on 'any significant environmental impacts'. Floorspace thresholds are given to provide an indication of significance and the need for environmental assessment.

The first 'key test' is the 'impact test' to assess the economic impact of a proposed retail development, including its cumulative impact. This has been regarded in the past as an essentially quantitative assessment leading to

Table 5.1 PPG6 – assessing new retail developments (the key tests)

	PPG6 paragraphs
The sequential approach and need	
Evidence of the sequential approach to site selection	1.9–1.11
The availability of suitable alternative sites	
The need or capacity for retail development	
Impact on development plan strategy	
Would a proposal undermine the strategy?	4.2
Impact on the vitality and viability of existing centres	
(the 'impact test')	
Define catchment area of proposed development	Annex B
Assess vitality and viability of centres in the catchment area	Figure 1
Quantitative assessment: analysis of trading impact by centre	4.13–4.15
• 'broad approach'	
• economic impact	
• cumulative impact	
Qualitative assessment: impact on vitality and viability of	
centres	4.3
• the risk to town centre strategy	
• the effect on future investment	
• changes to quality, attractiveness and role of centre	
• changes to physical condition of centre	
• the effect on range of services	
• increases in vacancies	
Accessibility	
Accessibility by choice of means of transport	4.6–4.8
• routes, frequency of services, etc.	
• the proportion of customers likely to arrive by different	
means	
Impact on travel and car use	
The likely effect on overall travel patterns and car use	4.9–4.11
• changes over the catchment area	
• opportunities for linked trips	
Environmental impact	
Floorspace thresholds:	4.19
Out-of-town developments 20,000 sq. m. gross	
Urban areas (new sites) 10,000 sq. m. gross	

estimates of percentage trade diversion from particular centres and residual turnover in these centres. The requirement for quantitative assessment remains and the advice is that 'in assessing impact it is essential that the local authority take a long-term view' (paragraph 4.4). While there is still a need for quantitative analysis, a greater emphasis is now being placed on the use of qualitative assessment. The impact of a new retail development on an existing centre will vary according to the ability of the centre to withstand some loss of trade. Therefore, it is important to judge the health of centres and PPG6's Figure 1 lists a range of indicators of the vitality and viability of centres which are shown in Table 3.1 in this book. Local authorities are advised to collect information regularly on these key indicators. The qualitative assessment will then show the impact of a proposal on the vitality and viability of centres, considering the factors listed in the table.

An increased emphasis on qualitative assessment should help to alleviate some of the concern that was expressed by local authorities in the survey about the lack of impartiality and objectivity in many retail impact studies (see Chapter 8). The estimates and assumptions that underlie quantitative analyses can lead to bias in favour of a proposal, and it is often difficult for a local authority to know how reliable the impact assessment really is, hence the growing use of independent audits of RIAs that are being commissioned by local authorities. Qualitative assessments, on the other hand, are much more easily understood and their accuracy can be readily established.

The survey of local authorities also highlighted the need for better retail information and local shopping surveys. Improvements in the methodology of RIA are pointless without corresponding improvements in the availability and quality of information about existing shopping provision and shopping patterns. PPG6 recognises the importance of retail surveys so that shopping policy and decisions on applications can be based on accurate and up-to-date information. The assistance of the private sector with such surveys is acknowledged, both on existing provision and the future capacity of shopping centres, and in collecting indicators of vitality and viability.

The combination of quantitative and qualitative assessment, then, provides the basis for the 'impact test' on which a proposal will initially be judged. Attention then turns to transport factors. A proposal may satisfy the impact test but fail on the grounds of accessibility or impact on travel and car use. Much of the advice in PPG6 is rooted in the message that 'town and district centres should be the preferred locations for developments that attract many trips' (para. 1.3) and that out-of-centre development is less acceptable because it encourages trips by car. The relevant paragraphs on accessibility and travel impact are noted in Table 5.1.

Technical requirements

Much of the criticism of the application of RIA has been due to weaknesses in the methodology for quantitative assessment. The main weaknesses have been pointed out in Chapter 4, and this section highlights the key areas in which improvements are required. There is a clear need for a better methodological framework for RIA. The early experience of RIA was largely based on shopping models, but there is little evidence of the use of models in current practice in RIA in Britain. Gravity models have so many drawbacks in methodology, data, forecasts and assumptions, and have been so frequently criticised by inspectors at inquiries, that there can no longer be any confidence in the use of such models in RIA. The framework put forward in this chapter does not rely on gravity-based modelling but uses an expenditure flow matrix which is a more reliable means of estimating the patterns of expenditure flows from where people live to where they shop.

The credibility of the application of RIA depends very much on the quality of the data and the validity of the assumptions made in the quantitative assessment. RIA is still an 'inherently indeterminate exercise' (Norris, 1992). Later in this chapter advice is given on how the problems of data and assumptions can be tackled to improve the accuracy of the impact predictions. The major criticism of RIAs revealed in the survey of local authorities is their lack of objectivity. Each party at a public inquiry, for instance, can manipulate impact methods to meet its own objectives. Inspectors have to weigh up the nature of the evidence and make their own judgements. The local authority survey showed a clear view from local authority planners that retail impact studies should be more independent/objective/impartial. But how can this be achieved?

It has been suggested that on receipt of an application for major retail development, the local planning authority could commission a RIA from a retail consultant represented on an approved panel of experts and charge the fee to the applicant (Inman, 1995). There are merits in this approach. Inman considers there are a number of advantages:

- The local planning authority would be the client rather than the applicant. The consultant would then be free of any obligation to the applicant.
- The local planning authority would have no need to commission a separate impact assessment.
- The application would be determined in a shorter period.

At present such a procedure is not in prospect and there was no reference in the Environment Committee's report to completely independent RIAs. However, the application of RIAs would be improved if there was a statutory requirement for retail impact studies to be conducted independently and to be demonstrably free of bias. Inman's suggestions might ensure impartiality but they would not necessarily mean an improvement in the quality of the assessment. The only safeguard open to local authorities at the moment in checking the quality of retail impact studies is to seek a second opinion by commissioning an independent review or audit of the applicant's study from another consultant. An independent review should not be a duplication of work on the original study, but should check the accuracy of the data, any errors or omissions in the assessment, and the validity of the assumptions and forecasts. It should comment on any bias in the assessment. The review should confirm the conclusions or raise questions about them and should comment on the significance of the predicted impact.

The research by CB Hillier Parker for the DETR (1998) highlights the need for a common methodology for assessing impact. CB Hillier Parker suggests a combined retail, economic and transportation evaluation (CREATE). Under this approach the conventional step-by-step approach is divided into two elements:

> first, the measured variables, which should be capable of quantification based on survey and other data to produce a model of the current situation (which should be agreed);

> second, assessment of the predicted changes against relevant criteria (which may not be agreed but should be capable of sensitivity testing).
>
> (DETR, 1998: 98)

The CREATE approach is shown in Figure 5.1. CB Hillier Parker considers that it provides a comprehensive checklist of criteria necessary to assess the impact of large foodstores. It is designed specifically to assess proposed major food store developments rather than other types of retail development but it could be adapted to meet the requirements of assessing comparison shopping developments. However, like other previous attempts to define a consistent and robust methodology for RIA, detailed guidance is not presented on the practical application of the approach.

The recommended approach

In this section a framework for RIA is presented which is practical and readily understood. The framework has been developed and refined in retail impact studies over several years. It is a very pragmatic approach to describing existing shopping behaviour and predicting the impact of a proposed retail development on shopping behaviour in the future, including trade diversion from existing centres. The framework concentrates on economic impact which remains the fundamental consideration in PPG6 for assessing the effect of a new retail development on the vitality and viability of shopping centres. However, qualitative impacts also need to be considered, as required by PPG6.

An outline of the recommended approach is shown in Figure 5.2. It incorporates all the elements of the assessment required by PPG6 and summarised in Table 5.1. These elements are described in detail in this chapter and in subsequent chapters.

The methodology of quantitative impact assessment is shown in more detail in Figure 5.3. The diagram illustrates all the inputs required and the

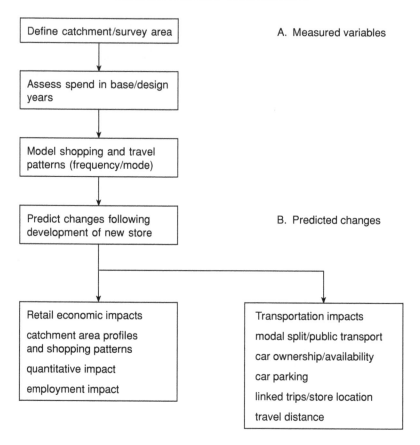

Figure 5.1 Combined retail, economic and transportation evaluation (CREATE)
Source: DETR/CB Hillier Parker.

steps involved in analysing trading impact, leading to conclusions on the interpretation of impact.

This framework is a marked improvement on the conventional step-by-step methodology of RIA described in Chapter 4 for the following reasons:

- the data inputs and assumptions shown in the flow diagrams are made explicit; there is no ambiguity about sources of data or projections
- the use of household survey data to estimate market shares is much more accurate than other methods of estimating turnovers
- the approach is not just concerned with knowing total expenditure in a sub-area (or isochrone) and total turnover of centres; the emphasis is on expenditure flows which are a close approximation to reality
- all expenditure is allocated in the retail sector, and the expenditure flows reflect shopping patterns for all centres in the catchment area

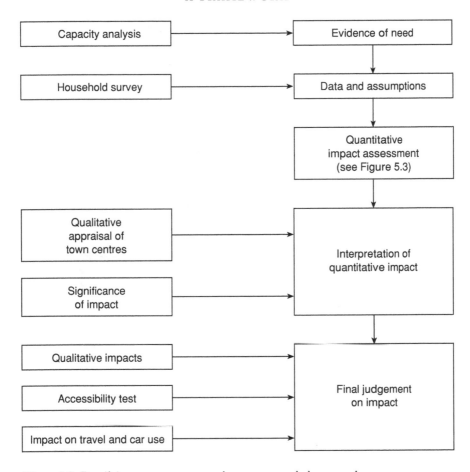

Figure 5.2 Retail impact assessment: the recommended approach

- leakage and inflows of expenditure are specifically included, and
- in assessing impact the framework replicates how shoppers themselves decide where to shop and how they will change their shopping patterns if a new store is developed.

It should again be stressed that the framework is not a shopping model. It does not rely on gravitation principles to allocate retail expenditure between zones or sub-areas and shopping centres. The allocation is based on observed shopping behaviour. The use of a computer spreadsheet for the distribution of expenditure also allows changes in inputs to be made very easily. In assessing the impact of proposals, different assumptions about turnover and trade draw can be made and the sensitivity of assumptions, e.g. on the turnover of the proposed development, can be tested.

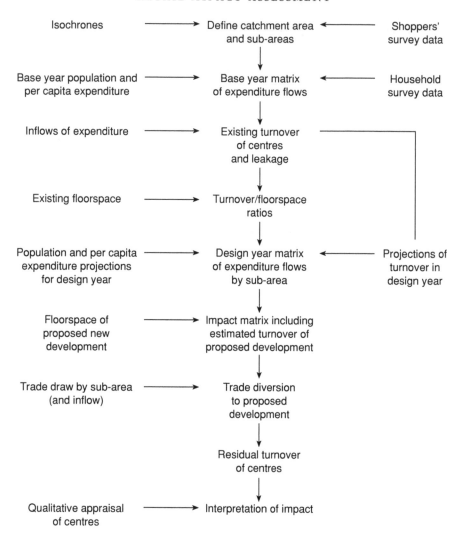

Figure 5.3 Methodology for quantitative assessment of retail impact

The final stage of the framework approach is the interpretation of impact. This must take into account the implications of trade diversion in the light of the qualitative appraisal of town centres outlined in Chapter 6 and the significance of impact in the context of the vitality and viability index of particular centres. This will enable a thorough analysis to be made of quantitative impact and its significance. It is then necessary to make a final judgement about the impact of a proposed development by looking at other qualitative factors (the impacts on the vitality and viability of centres listed in para. 4.3 of PPG6) and the 'sustainability' tests in PPG6 which are those concerned with accessibility and the impact on travel and car use.

This approach satisfies the need for improvements in RIA methodology, and meets the policy requirements of PPG6 in terms of taking a broad approach, a long-term view, and facilitating agreement on assumptions between the parties at inquiries. PPG6 could have been more explicit about the practical application of RIA, rather than just setting out general principles. The best practice advice in this chapter is intended to fill this gap. The framework approach has been tested in practice and shown to be a reliable method of predicting the impact of new retail developments. This approach, combined with qualitative inputs and the other tests embodied in PPG6, is a positive step forward in making informed decisions on proposals for new retail developments.

Capacity analysis and quantitative need

A growth in capacity in the retail system is the first indication of retail demand. Capacity analysis uses the same base data as RIA on retail expenditure. It takes expenditure growth from the base year over a long period, usually 10 years, over a wide area. The growth of retail expenditure represents turnover potential which can be converted to floorspace capacity by applying turnover/floorspace ratios. The net floorspace requirement needs to take account of any commitments and, particularly in the case of comparison shopping, the increase in floorspace efficiency of existing shops. The reuse of vacant floorspace should also be considered.

Capacity analysis is most useful for assessing the scope for additional comparison goods floorspace but for convenience goods it is less useful. If there is little capacity for additional floorspace, which is usually the case for convenience goods, a capacity analysis can confirm that any major new development would result in an impact on trade in existing centres. In the case of comparison goods, care must be taken in estimating the floorspace capacity which may arise from an increase in turnover potential. Turnover/floorspace ratios for 'town centre' comparison goods retailing are much higher than for bulky non-food goods of the type commonly sold in retail warehouses.

The methodology of capacity analysis has some major limitations. Above all, it is simplistic. The analysis takes no account of the dynamics of the retail system, particularly for changes to occur in shopping patterns as new developments take place. It assumes that the retail system is in an equilibrium state in the base year and that the turnover of existing shops will increase very gradually over the forecast period. In reality, where existing shops are under-trading there will be scope for a more rapid growth of turnover which will therefore reduce the capacity for new development. Where existing shops are over-trading there will be less opportunity for growth in floorspace efficiency and so the potential for new floorspace could be greater. Another issue to be considered is that retention levels and leakage are not explicitly considered. Retention levels are a function of the degree of self-containment of a catchment area, which is influenced by the way the area is defined geographically. A catchment area which has a net leakage of trade may have the potential to increase its trade retention if a new development takes place.

Chapter 4 referred to the two components of need: *quantitative need*, which is the capacity of a catchment area to support additional retail development without harm to existing centres, and *qualitative need*, which is concerned with the choice of shopping facilities (Holt, 1998).

This section focuses on quantitative need. (Qualitative need is discussed in Chapter 6.) PPG6, para. 1.10, requires the local authorities to consider the need for new development in drawing up their development plans. It states that 'if there is no need or capacity for further developments, there will be no need to identify additional sites in the town'. However, need is not identified as one of the key considerations in para. 1.16 of PPG6 in respect of out-of-centre developments nor in para. 4.13 in respect of all retail developments in excess of 2,500 square metres gross floorspace (Holt, 1998).

The ambiguity surrounding need in PPG6 has led to some important High Court cases. The most significant was the High Court judgment in 1998 on a legal challenge by Somerfield Stores into a decision by Hambleton District Council to grant planning permission for a Safeway superstore in Northallerton. This was a landmark judgment in two key areas:

- that need could be a material consideration in certain circumstances, e.g. if there is evidence of quantitative or qualitative need which supports a development, but a developer does not have to demonstrate need for the development
- that a proposal which would not sustain and enhance a town centre could still be acceptable if it meets the sequential test and other key tests in PPG6 (Arnold, 1998).

The Hambleton judgment was short-lived. In February 1999 the Planning Minister, Richard Caborn, made a statement intended to clarify the guidance in PPG6 in the light of issues raised in this case and other recent litigation concerning government policy on retail development. The minister said:

> Proposals which would be located at an edge-of-centre or out-of-centre location . . . should be required to demonstrate both the need for additional facilities and that a sequential approach has been applied in selecting the location or site.
>
> In the context of PPG6 and this additional guidance, the requirement to demonstrate *need* should not be regarded as being fulfilled simply by showing that there is capacity (in physical terms) or

demand (in terms of available expenditure within the proposal's catchment area) for the proposed development. Whilst the existence of capacity or demand may form part of the demonstration of need, the significance in any particular case of the factors which may show need will be a matter for the decision maker.

(DETR, 1999a)

Government policy on need, therefore, has been clarified but there is still no definition of it. The issue remains – how is need defined and by whom? (Goddard, 1999). What is clear is that quantitative need comprises the following elements:

- economic capacity in terms of demand arising from expenditure growth within the catchment area of a proposal
- leakage of trade from an area, which suggests a lack of provision to meet the needs of shoppers
- retailer requirements – demand by retailers for representation in a particular centre, and the potential for competition and innovation.

Data and assumptions

Serious data problems beset any quantitative assessment of retail impact. Many writers over the years have pointed to the overriding need for a better and more comprehensive retail information base. Particular weaknesses are the availability and reliability of data on shopping floorspace and turnover. The local authority survey in Chapter 8 also clearly shows the need for better/more up-to-date retail statistics at the local level.

Although there are serious problems with data, with intelligent use of the data that is available, the data quality need not be a handicap to undertaking sound impact assessments. There are a number of key areas on which decisions have to be made on data and assumptions and these are covered in this section.

Household surveys

The conventional methodology of RIA, described in Chapter 4, involves defining catchment areas by means of isochrones and preferably zones or sub-areas. Sub-areas allow a more accurate geographical representation of shopping patterns. Where household survey data are available, the use of sub-areas is particularly important in helping to assess expenditure flows. However, the sub-areas must be defined to represent a meaningful geographical pattern of shopping trips.

The use of household surveys is recommended because recent experience of undertaking RIAs shows that it is the only way to build up an accurate picture of existing shopping patterns. The study area selected for a household survey need not represent the defined catchment area of a centre but it should be sufficiently large to comprise the primary catchment area and other areas from which a significant proportion of trips will go to that centre. An interview survey of shoppers in a centre can be used to show where shoppers live and so help define the full extent of the catchment area. But if there is no shoppers' survey data, analysis of the household survey results by sub-area/postcode will still enable the catchment area to be defined accurately.

The nature of the questions to be asked in a household survey will vary according to the type of retail study being carried out, but generally the key questions will be concerned with:

- the store mostly used for main food shopping
- the mode of travel for main food shopping
- the shops used for top-up food shopping
- the proportion of food shopping spending that goes on the 'main shop'
- the centres used for non-food shopping (different categories of comparison goods)
- the centre mostly used for non-food shopping
- whether shoppers make linked trips when doing main food shopping.

Some attitudinal questions may also be asked, e.g. how satisfied people are with a particular centre, their specific likes and dislikes about aspects of the centre (the range and quality of shops, car parking, the shopping environment, etc.) and ways in which it might be improved.

The normal rules of questionnaire design should apply, such as wording questions so that they do not lead to biased or ambiguous answers. The sample for a household interview survey should be large enough to be statistically reliable. The sample size depends on two factors: the degree of accuracy required, and the extent to which there is variation in the population.

There are several things to note about the relationship between sample size and accuracy (de Vaus, 1996):

- When dealing with small samples a small increase in sample size can lead to a substantial increase in accuracy. Beyond a certain point the cost of increasing the sample size is not worth it in terms of extra precision. A sample size of 2,000 is considered to be sufficiently accurate in any situation since beyond this point the extra cost does not lead to any significant gain in accuracy.
- The size of the population from which the sample is drawn is largely irrelevant for the accuracy of the sample. It is the absolute size of the sample that is important.

- To analyse sub-groups within the sample, it should be sufficiently large so that the smallest sub-group has at least 50 cases.

The final sample size will be a compromise between cost, accuracy and ensuring sufficient numbers for a meaningful sub-group analysis. The researcher must decide what use is to be made of the results and what precision is required (Moser and Kalton, 1993). In most cases a sample of 400 households would be acceptable, producing a margin of error of 5.0 per cent at the 95 per cent confidence level. The researcher's choice of confidence level will depend on what is most important – a narrow range of error or a high probability of being correct. But a 95 per cent confidence level is used in most instances (Moser and Kalton, 1993). A sample of 500 would have a margin of error of only 4.5 per cent. Increasing the sample size any further is not necessary for statistical accuracy and it would increase the survey costs beyond what is reasonable.

The exception is that for very large populations; it may be justifiable to use a larger sample of say 1,000 households to increase the statistical coverage of the population. A sample size of 1,000 has a margin of error of less than 3 per cent (de Vaus, 1996). It is also not necessary to have a sample of more then 10 per cent of households, so that in a small catchment area of say 4,000 households, the sample size does not need to be more than 400.

Household surveys should wherever possible be conducted as in-home, face-to-face interviews by trained, professional interviewers, and the number of interviews should be quota-controlled by sub-area. In-home surveys produce considerably better quality information than telephone surveys which have an inherent bias because a significant proportion of households are without a telephone or are ex-directory. Telephone interviews are also subject to respondent suspicion about telephone selling, and they are not significantly cheaper to carry out.

Expenditure

In recommending best practice on retail expenditure data and assumptions, the key decisions are: whether to use a goods-based or business-based definition of expenditure and what is the most appropriate growth rate to use for expenditure projections. It is advisable to use expenditure data from The Data Consultancy (now MapInfo Ltd), whether in the form of national expenditure estimates published in The Data Consultancy's annual information briefs, or local estimates which can be specially ordered as Illumine reports for local authority areas or other small areas. Both these sources of data are available as goods-based and business-based expenditure. There is at present no overall consensus among practitioners about whether the goods-based or business-based definition is preferable. The advocates of business-based expenditure claim that it is better because in many proposals for superstores the operator is not known and so the arrangement of food and non-food goods floorspace cannot be determined and all company average turnovers derived by *Retail Rankings* and other organisations based on company annual reports are business-based.

The difficulty with the business-based definition is that the term 'business' does not relate to individual shops but to the whole company, e.g. Co-ops are defined as convenience businesses and their department stores are treated as part of that convenience business. On the other hand, Marks and Spencer food halls are classed as part of the company's business as a 'comparison mixed' retailer.

According to URPI (1995) neither goods-based nor business-based analysis can be authoritatively pronounced better, since each has both strengths and weaknesses. URPI states that in general the expenditure data are likely to be more accurate at a local level on a *goods* basis, but for convenience shopping it is becoming increasingly common to use business-based expenditure.

The conclusion for best practice is that in most cases goods-based expenditure and turnover should be used. A goods-based definition should certainly be used in any analysis of comparison shopping, partly to avoid the Marks and Spencer food hall problem mentioned earlier, and because any analysis of retail warehousing should use the URPI data on bulky goods expenditure which is by goods type (see below). It is true that floorspace and turnover/floorspace ratios are more easily obtained on a business basis, but it is good practice in any retail study to make an accurate survey or estimate of shopping floorspace and this can be done by goods type if the effort is made to adjust floorspace for food and non-food goods in larger stores. The most clear justification for using a business basis is in the case of a superstore proposal in a conurbation with numerous existing superstores and variety stores where the adjustment of floorspace may be too onerous. However, the problems of unreliability of convenience expenditure trends on a goods-basis, referred to earlier, makes it advisable to use the business-basis for the assessment of proposals for all convenience shopping developments.

The other key decision to make on retail expenditure concerns the annual rate of expenditure growth. Per capita expenditure projections normally use local figures of per capita expenditure per annum as the base, to which is applied national growth rates published by The Data Consultancy. These are updated annually for convenience and comparison shopping (both by goods type and business type) and are based on trends over different time periods. The Data Consultancy recommends the most appropriate growth trend to use. Its latest advice is to adopt growth rates by goods type based on long-term or ultra long-term growth trends for comparison shopping, and by business type based on either long-term or short-term growth trends for convenience shopping.

The Data Consultancy has also analysed expenditure growth in bulky goods (DIY/hardware, carpets and furniture, and electrical goods) in Information

Brief 98/3. The application of these more detailed trends is strongly recommended when dealing with retail warehouse developments, but these trends should only be applied for short-term projections.

Turnover

Estimating the turnover of existing centres by applying typical turnover/ floorspace ratios to estimates of retail floorspace is acceptable in cases where there is no alternative data, but it is better to use market shares for individual centres/stores if data are available from a household survey. The Drivers Jonas research (1992) advised that estimates of turnover should be based on local survey information or controlled by knowledge of available expenditure in the local retail system. The market share approach involves taking the survey responses to show the percentage of people who mainly shop at named centres or stores for different types of goods.

Empirical experience of using this market share approach shows that it is very reliable and produces turnover estimates that can reasonably be verified by other sources such as evidence of trading performance. But in some situations the survey data may not accurately represent market shares for the following reasons:

- The data are for shopping trips, not expenditure. The average spending per trip may vary significantly between different stores, e.g a superstore and a smaller supermarket.
- The survey may under-represent the extent to which shoppers use smaller shops, e.g. specialist food shops, because they only name the shop(s) they use most.

Therefore the figures should not be accepted at face value. The turnovers per square metre of the main stores and other shops should be checked carefully to see if they look reasonable, and professional judgement should be applied to make adjustments where necessary.

In estimating the future turnover of centres in the design year for the purposes of the impact assessment, it is conventional to assume that some increase will take place in these turnovers. Evidence of the rate of increase is lacking but the growth of floorspace efficiency is greater for comparison-goods shops than it is for convenience-goods shops. Floorspace efficiency for convenience-goods shops increased significantly in the past with the 'self-service revolution' and now it is increasing only slowly. In terms of convenience goods, the turnover of existing shops can be expected to increase at the same rate as the growth of expenditure. For comparison goods it is often assumed that the floorspace efficiency of existing shops will increase at a rate of 1.5 per cent per annum, but there is a lack of evidence to support this assumption (Thorpe, 1994). In the context of an increase in comparison-goods expenditure of 3 to 4 per cent per annum in real terms, a growth of floorspace efficiency of around 2 per cent per annum is probably more realistic. Therefore, for comparison goods, the base year turnovers should be increased by 1.5 to 2 per cent per annum over the period up to the design year. This is important in assessing the residual turnover of centres (see later section, 'Quantitative impact').

The turnover of a proposed store is normally estimated by multiplying its net floorspace by an appropriate figure of turnover per square metre (including VAT but excluding petrol sales in the case of superstores). Company averages are often used as the basis for estimating the turnover per square metre but they are a poor guide to the likely turnover of an individual store. There can be wide variations in the performance of stores depending on location, degree of competition, etc. In practice the likely turnover of, for instance, a proposed superstore should take the company average as the starting point, then adjust it up or down according to judgement about the trading position of the centre or other similar stores in the area.

Expenditure flows

The assessment framework is based on the analysis of expenditure flows for a retail sector (convenience or comparison). An expenditure flow matrix is built up for the base year using household survey information which shows the pattern of shopping between sub-areas and centres/stores. The base year matrix is set up as a computer spreadsheet in which the rows are sub-areas and the columns are stores or centres. The final column is total expenditure (convenience or comparison) and the penultimate column is leakage out of the catchment area. The final row is turnover of each centre and the penultimate row is inflow of expenditure into the catchment area. All expenditure is allocated using this framework, so it is necessary to have a column for local centres or shops not included in the other columns.

Expenditure from each sub-area is allocated between centres based on the market shares shown by the household survey. For the convenience sector it is preferable to produce a matrix for main food shopping and top-up shopping separately, then add these together for all convenience goods. For comparison shopping, if data are available, it is possible to produce separate matrices for each category of non-food goods then aggregate them for all comparison goods. The turnover of a particular centre is then the summation of the expenditures from different sub-areas to that centre. As was noted earlier, in some situations it may be necessary to adjust the market shares from the household survey to counteract any under-representation of smaller shops or centres, and this should be done before completing the base year matrix.

This matrix forms the basis for predicting retail impact. To test the impact of a proposed development it is necessary to set up an impact matrix for the design year, projecting expenditure forward say two to three years. Projections are made of expenditure for a particular sector by sub-area. Estimates are then made of the turnover of the proposed development and its trade draw by sub-area (including inflows). The trade draw to that development from each sub-area is subtracted from total expenditure for the sub-area, and the remainder is redistributed among centres. Reasoned judgement must be used so that the method is able to stand up to scrutiny from an independent review or challenge from other parties at an inquiry.

The predicted turnovers of centres are compared with the projected turnovers in the design year to assess trade diversion. In the impact matrix additional rows are included in the spreadsheet for trade diversion, percentage impact and residual turnover per square metre. To test cumulative impact, the impact matrix is set up in the same way for the design year, and the estimated turnovers of the proposed developments are input to the matrix. If there are two or three proposals it is normal practice to reduce their turnovers by an appropriate percentage, as noted in Chapter 4. The trade draws to each proposal are then calculated by sub-area and the residual expenditure from each sub-area is redistributed to existing centres. Finally the predicted and existing turnovers are compared to assess trade diversion.

Quantitative impact

Trading impact

Estimating the trading impact of a proposed development should be done in three stages: making assumptions about trade draw, calculating the percentage trade diversion from existing centres, and assessing the residual turnover of centres.

Trade draw

Estimating the proportions of trade to a proposed development that would be drawn from each isochrone/sub-area is a difficult but very important element of any RIA. Evidence can be derived from existing stores but it is necessary to take account of the distribution of population in the catchment area and the pattern of existing shopping provision in the area. It is usually assumed that up to 10 per cent of trade will be drawn from outside the primary catchment area and that the percentage of trade from each isochrone will decline with distance from the store, the same principle that underlies gravity-based shopping models. This is a key area for possible agreement between the parties at a public inquiry because even small differences in the assumptions can have a significant bearing on the final impact figures.

Trade diversion

The assessment of trade diversion requires the amount of trade predicted to be lost from each centre to be expressed as a percentage of that centre's turnover. The denominator is the centre's projected turnover in the design year. The trade diversion from particular centres can simply be estimated subjectively but this is inadequate and would be open to close scrutiny at an inquiry. It can also be assumed to be pro rata to existing shopping patterns, i.e. a centre will lose trade in proportion to the amount of trade it draws from each isochrone or sub-area. This is also not a totally acceptable method because in reality stores tend to compete like with like. Therefore, the recommended approach is the subjective allocation of residual expenditure from each sub-area to each centre based on the types of centre/store most likely to compete with the proposed development. This is a close approximation of how shoppers will change their shopping behaviour after a new store opens.

Residual turnover

Retail impact should not be assessed in terms of percentage trade diversion only. It is also essential to assess the residual turnover of centres, expressed as turnover per square metre in the centres which are predicted to experience trade diversion in the design year. Residual turnover per square metre can be compared with company averages or typical turnover/floorspace ratios for different types of centres, but it is more reliable to compare residual turnover with the minimum level of viability of a store or centre – the level below which closures are likely for a particular type of store.

Sensitivity of assumptions

Because of uncertainties inherent in the assumptions and projections used in RIA, it is advisable in any impact study to test the sensitivity of the assumptions. It is necessary to build in an allowance for variations in the assumptions and errors in forecasting. Experience shows that the main factors in the assessment which are subject to variability are the level and rate of growth of per capita expenditure, the predicted turnover of the proposed development, and the amount of clawback of leakage.

The choice between goods-based and business-based expenditure has already been discussed. Where there is some dispute over the use of goods-based or business-based figures, or where there are good reasons for considering a range of expenditure growth projections, the goods-based and business-based figures can be regarded as a range with goods type as the lower limit and business type as the upper limit. The sensitivity of expenditure growth can also be tested by adopting different rates of growth from those recommended by The Data Consultancy. The projections or per capita expenditure are extrapolations of trends over different time periods to which a regression line has been fitted. The recommended trends are those with the highest correlations, but there may be circumstances in which an alternative projection may be preferred. For example, it may be considered inappropriate to apply national growth rates for expenditure in an area which is depressed economically. Therefore, it is common to take alternative trend projections as a basis for a range of expenditure growth, typically The Data Consultancy long-term and ultra long-term trends, or a Data Consultancy trend line and some other assumed growth rate which is lower to produce a lower limit for the projections. Even if no deliberate choice of a range is made, it is important to bear in mind the uncertainty about the rate of growth of retail expenditure. All The Data Consultancy's expenditure projections are subject to 95 per cent confidence limits, and these should be reflected in the interpretation of retail impact.

The second key factor in testing the sensitivity of an impact assessment is uncertainty about the turnover of the proposed development. Its predicted turnover may be based on company averages but it is necessary to allow for local variations above or below the average level. A range of predicted turnovers per square metre may be used to make the analysis more robust. The turnover of a foodstore will depend to a large extent on who the operator will be. If there is a named operator, the turnover should reflect its typical trading performance.

The other key factor which is relevant to sensitivity analysis is clawback of leakage to the proposed development. The extent to which new stores will claw back trade currently lost from the catchment area can be a matter of contention between developers and local authorities. Expectations as to the prospect for clawback of trade must be realistic and reflect the centre's position in the retail hierarchy relative to larger centres with overlapping catchments (CB Hillier Parker, 1998). CB Hillier Parker's analysis in relation to foodstores is that the amount of clawback will depend on:

- the size and accessibility of the store
- its location (out-of-centre or edge-of-centre)
- the nature of the catchment area (the range and proximity of competing stores), and whether any other non-central foodstores have been developed in the locality.

Summary

There is a clear need for advice on best practice on the application of RIA in Britain. Evidence of the need for improvement is shown by criticisms by the local authorities of retail impact studies, the recommendations of the House of Commons Environment Committee, comments by retailers and practitioners in the field, and the sceptical attitude of planning inspectors. Approaches to RIA have not kept pace with changes in the policy context in the last decade, and there needs to be a much more pragmatic approach to retail impact issues.

PPG6 sets out the key tests for assessing retail developments – the 'impact test', and the 'sustainability tests' on accessibility and travel/car use. The guidance also introduced a sequential approach to site selection and requires an assessment of need or capacity for retail development. PPG6 places an emphasis on qualitative as well as quantitative factors in assessing impact on the vitality and viability of centres.

There is a need for a methodological framework for RIA which is independent and objective. The only safeguard open to the local authorities at the moment is to commission independent reviews or audits of retail impact studies. In this chapter a framework has been recommended for RIA based on expenditure flows which is practical and readily understood. It has been developed and refined over several years and shown to be a reliable method of predicting the impact of new retail developments. The framework is a marked improvement on the conventional step-by-step methodology of RIA.

Government policy in PPG6 and ministerial statements require proposed new retail developments to be justified by evidence of capacity and need. Capacity analysis shows the quantitative requirement for additional shopping floorspace. The methodology has some major limitations and it must be applied cautiously and intelligently. The government has made it clear that proposals for new retail development outside town centres must be supported by evidence of need. Quantitative need is concerned with economic capacity or demand, leakage and retailer requirements.

Important decisions have to be made about data and assumptions. Particularly important is the use of household surveys to provide accurate data on existing shopping patterns. Household surveys also enable turnover to be estimated based on market shares rather than making estimates of turnover per square metre based on company averages. Retail expenditure and turnover can be defined on a goods-basis or a business-basis, depending on the nature of the catchment area. Comparison shopping is normally assessed on a goods-basis, and convenience shopping on a business-basis.

A base-year expenditure flow matrix should be set up as a computer spreadsheet, representing flows of expenditure from sub-areas to shopping centres. In order to predict retail impact, expenditure is projected for the design year and an impact matrix devised which assesses the effect of a proposed development on shopping patterns. It can also test the cumulative impact. In estimating the trading impact it is necessary to consider the trade draw from different parts of the catchment area, the percentage trade diversion from centres, and the residual turnover of centres. The sensitivity of assumptions must be tested, especially on future expenditure, the predicted turnover of a proposed development, and the amount of clawback of leakage.

6

QUALITATIVE FACTORS

It is no longer adequate to assess retail impact simply in quantitative terms. PPG6 refers to several qualitative factors which must also be considered, and these are listed in Table 5.1 in the previous chapter. Figure 5.2 shows how these qualitative factors form an integral part of the recommended framework approach to RIA. The essential qualitative factors are:

- health check appraisals of town centres
- qualitative need
- the sequential approach
- accessibility
- the impact on travel and car use
- environmental impact.

These factors are considered in this chapter, together with the interpretation of the significance of retail impact.

The qualitative factors to be considered in assessing applications which may have an impact on the vitality and viability of centres are listed in paragraph 4.3 of PPG6. They are:

- the extent to which developments would put at risk the strategy for the town centre, taking account of progress being made on its implementation
- the likely effect on future private sector investment needed to safeguard the vitality and viability of that centre
- the changes to the quality, attractiveness and character of the centre, and to its role in the economic and social life of the community
- the changes to the physical condition of the centre
- the changes to the range of services that the centre will continue to provide, and
- the likely increases in the number of vacant properties in the primary retail area.

Town centre health check appraisals

The concept of vitality and viability has become the cornerstone of government policy on sustaining and enhancing town centres. PPG6 emphasises
the need to assess the vitality and viability of town centres using health
checks and Figure 1 in PPG6 lists a number of indicators which are recommended as the basis for undertaking health checks. A thorough appraisal of
centres which may be affected by retail impact is essential. Advice on RIA
by Norris and Jones (1993) advocated that the approach should move away
from the quantitative assessment of impact to a qualitative centre-based
survey approach which evaluates the strengths and weaknesses of centres and
their ability to withstand new competition.

In this section a framework is set out for the appraisal of the health of
town centres which has been shown from empirical testing over a period of
several years to be a reliable means of qualitative assessment. It begins with
the premise that the indictors listed in PPG6, and shown in Table 3.1,
should be used in any health check appraisal. They have been revised since
the 1993 version of the guidance and they have been tested in numerous
retail impact studies and at public enquiries. However, they are fairly broad
indicators which do not lend themselves to systematic application. PPG6 is
not prescriptive about these indicators. No advice is given on how they
should be applied in practice, and they are too subjective to be a reliable
guide to the health of a town centre. For example, how can general terms
such as 'retailer representation' or 'accessibility' be judged meaningfully in a
way which can act as a measure of vitality and viability? Several consultants
have devised their own health check approaches but they can be excessively
detailed and expensive.

To overcome these problems of assessing the health of a centre the indicators in PPG6 were compared with the more detailed 'factors' of vitality and
viability suggested in the URBED research report, 'Vital and viable town
centres', commissioned by the Department of the Environment (URBED,
1994). There are many similarities in the nature of the indicators and factors, but URBED grouped its factors into specific elements of attraction,
accessibility and amenity. The PPG6 indicators and corresponding URBED
factors have been combined to produce an integrated checklist of factors that
are relevant to the appraisal of a town centre. The relevant factors were listed
against the PPG6 indicators to provide a total of 42 factors that can be
assessed for any town centre. This checklist is shown on the appraisal sheet
in Table 6.1.

The appraisal framework is based on a scoring system using a five-point
scale of 1 = very poor, 2 = poor, 3 = fair, 4 = good, and 5 = very good. Each
factor is given an assessment score from 1 to 5. Most of the data needed for
the appraisal can be obtained from a comprehensive town centre survey of
the use of retail and commercial properties and their physical appearance. The

Table 6.1 Town centre health check appraisal sheet

Indicator	Factor	Assessment*
Diversity of uses	Number and range of shops Financial and professional services Business and office premises Cafés and restaurants Pubs and clubs Cultural attractions/community facilities Sports and leisure facilities	
Retailer representation	Number of multiple retailers Variety of specialist/independent shops Existence and quality of market Availability of food shopping Availability of enclosed shopping Retail opening hours Evidence of recent investment by retailers Retailer demand Presence of charity shops Presence of low quality discount shops	
Vacant properties	Vacancy rate Vacant floorspace Effect of vacant premises on the centre	
Commercial performance	Rental values Shopping centre yield	
Pedestrian flows	Volume of pedestrian flow	
Accessibility	Ease of movement for pedestrians Ease of movement for cyclists General pedestrian environment Ease of movement by car Car parking Quality of public transport Ease of movement by public transport Ease of movement for the less mobile	
Customer views and behaviour	Satisfaction with the centre Need for improvements Leakage of trade to other centres	
Safety and security	Feeling of security Perception of safety outside shopping hours Availability of CCTV	
Environmental quality	Physical appearance of properties Overall cleanliness Quality of buildings Quality of open spaces/ landscaping Availability and condition of toilets	

* 1 = very poor; 2 = poor; 3 = fair; 4 = good; 5 = very good.

survey information can be analysed to derive scores for most of the factors corresponding to the indicators of diversity of uses, retailer representation, and vacant properties. Other factors relating to pedestrian flows, accessibility, safety and security, and environmental quality can be obtained by observation of the centre generally and from information available from the local authority. Two of the indicators require information that must be obtained from other sources: commercial performance requires data on rental values and shopping centre yields which were discussed in Chapter 3, and customer views and behaviour require data from household or shopper interview surveys.

To illustrate how the factors can be applied in practice, examples are given below of the way in which scores are attributed to particular factors. Taking the example of the first indicator, diversity of uses, one of the most important factors is the number and range of shops. A relatively large number of shops will tend to produce a relatively high score because the larger a centre is, the more attractive it is likely to be to shoppers. Smaller centres will not necessarily imply a lower score because the role of the centre must be taken into account. A district centre with relatively few shops may still be regarded as 'good' if it has a relatively large number of shops for its role. But the absolute number of shops is not the only measure of attraction. The range of shops is also important. A centre which has a narrow range of shops, say with an emphasis on convenience shopping, and a relative lack of comparison outlets, will tend to have a lower attraction than a centre with a good range of both convenience and comparison outlets in which shoppers are offered more variety of goods. It is, therefore, the quality of the opportunity for purchasing goods rather than just the size of the centre which is significant. This will influence whether a score is given which is better or worse than average. A deficiency of certain types of shops is likely to produce a 'poor' scoring, whereas a 'good' score would reflect a centre which offers a greater range of opportunity.

In the case of the second indicator, retailer representation, it should be noted that for two factors – presence of charity shops and presence of low quality discount shops – a 'negative' scoring is to be applied. The presence of charity shops, for example, should be regarded as a negative factor because it suggests that the centre is unable to support a full range of viable outlets paying market rents. Charity shops reflect 'concealed vacancies', i.e. units which would otherwise be vacant if it were not for the presence of the charity organisations. A perceived large number of charity shops would suggest a relatively poor scoring whereas the absence of charity shops would suggest a relatively good scoring. The same principle applies to low quality discount shops, the presence of which can usually be taken as a reflection of a centre which is performing at below average levels of trading, and therefore would suggest a 'poor' scoring.

In the scoring system no attempt is made to weight individual factors. It would be extremely difficult to do so and there could be no sound justification for the weightings used. But in fact the scoring system tends to give greatest weight to the indicators of diversity of uses, retailer representation and accessibility because they have the largest number of individual component factors. These are the indicators which are generally regarded as being the most important measures of vitality and viability. Conversely, least weight is given to indicators of pedestrian flow and commercial performance which are generally regarded as being of lesser importance. Other indicators such as vacant properties, customer views and behaviour, and environmental quality lie between these extremes. This is an accurate reflection of the relative importance of different indicators in assessing the vitality and viability of centres.

Comments have already been made in Chapter 3 about the use of indicators of commercial performance, particularly shopping centre yields. Great care is needed in the use of such data. For small centres data will probably not be available on rental values or yields. Another indicator, accessibility, is discussed later in this chapter. The final indicator in Table 3.1, on environmental quality, is an important element of any town centre appraisal survey. It requires observation of the physical appearance of retail and commercial properties in the centre, scoring them again on a five-point scale from 1 = very poor to 5 = very good. The scoring of individual properties is a skilled exercise based on judgements about the condition of properties, the standard of maintenance and decoration, and the image which premises present to potential customers. Other factors of environmental quality are also based on the town centre appraisal survey.

It is not necessary for all the factors to be scored; not all of them will be applicable to every centre, or in some cases data will not be available. An average score is calculated just for those factors that do apply. The end-product of the appraisal, then, is an overall score or vitality and viability index for a centre. Experience shows that the vitality and viability index will tend to range from below 2.5 for a centre which is performing badly with a low level of vitality and viability to 4.0 or more for a centre which is performing well with a high level of vitality and viability. An index of 3.0 indicates an average or fair performance. The approach has been applied successfully to a large number of shopping centres in the north of England since 1996. It has been refined since the latest version of PPG6 but appraisals carried out previously have been revised using the same framework and consistent results have been obtained. Most of the centres assessed were found to have overall scores of between 2.5 and 4.0. Examples of the vitality and viability indices for some of the centres which have been assessed are shown in Figure 6.1.

The appraisal framework in Table 6.1 is a subjective but systematic approach to assessing the vitality and viability of centres. It is relatively simple to use in practice and it provides very reliable results. The approach allows comparisons between centres, but such comparisons should be made

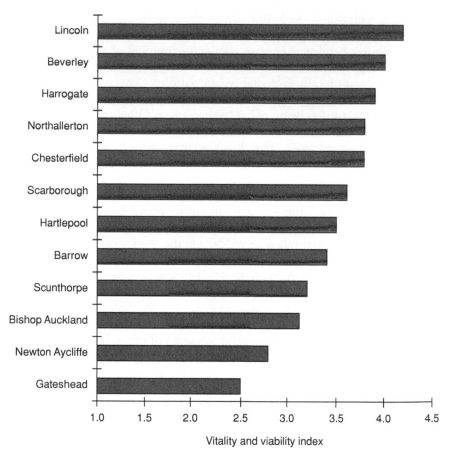

Figure 6.1 Vitality and viability indices

for centres which are similar in size and function, a process sometimes known as 'benchmarking'. The appraisal is a snapshot but by repeating the process over a period of time (annually or biennially), changes in the vitality and viability of a centre can be monitored. The approach is considered to be a significant improvement on other methods of carrying out health checks. It satisfies current government policy guidance, and the best practice advice in this book is that it should be applied in all retail impact studies for all centres which might be affected by a proposed development.

Qualitative need

Need represents a requirement for additional shopping floorspace. In Chapter 4 brief mention was made of qualitative need, defined as a geographical gap in shopping provision or a deficiency in the quality of provision. Qualitative need is not defined in PPG6 but the statement by the Planning Minister, Richard Caborn, in February 1999 makes it clear that the significance of factors which may show need will depend on local circumstances. The minister made a further statement on need in Parliament in June 1999 when he said:

> Developers must demonstrate that need does not simply mean an assertion by the developer that there is market demand. It means that the local planning authority must consider the wider needs of the community as well as the market demand . . . If the local planning authority is satisfied that a need exists, it must also be satisfied that the sequential test has been applied in selecting the site. Even then, the local authority must also consider whether there will be an adverse impact on the existing centre before it allows the proposal to go forward. We have tried to clarify the word 'need'. It will be for the local planning authority to make the decisions.
>
> (*Hansard*, 1999)

In assessing need it is necessary to examine:

- the nature and quality of existing shopping provision – are there qualitative deficiencies in types of provision, by sector or geographically?
- indications of over-trading – is there evidence that existing shops are not meeting the available demand, e.g. 'trolley congestion' and inadequate car parking?
- consumer demand – do surveys of customer views show that new development would meet the requirements of the local community?

A report by the English Historic Towns Forum (1997) gives guidance on assessing the qualitative need for new retail floorspace. It recommends that a number of qualitative factors should be considered, as follows:

- are the current facilities outdated?
- are better facilities offered by competing centres?
- is there a leakage of trade from the catchment area?
- are the existing facilities cramped and overcrowded?
- are there gaps in the types of stores?
- to what extent is there an unserved catchment population?

A further matter, not often included in RIAs, is the 'no development scenario'. Impact studies should also address the consequences of a proposed development not proceeding. The issues arising from unmet need if there is a lack of new retail development include:

- the possible decline in established centres through lack of investment
- the increasing volume of shopping trips to competing destinations
- the lack of maintenance of an attractive and lively shopping environment (English Historic Towns Forum, 1997).

Most of the qualitative factors listed in paragraph 4.3 of PPG6 are factors concerned with qualitative need, such as investment in town centres, the quality of a centre, its physical condition and the services it provides. The advice on best practice in this respect is to apply subjective judgement, based on the appraisal of the health of a town centre, to assess if there are deficiencies in the type and quality of existing shopping provision that could be met by new retail development.

The sequential approach

PPG6 places emphasis on the sequential approach to selecting sites for retail development and other town centre uses. The origins of the sequential approach lie in PPG13 which states that where suitable central locations are not available, edge-of-centre sites close enough to be 'readily accessible by foot from the centre and which can be served by a variety of means of transport' should be chosen. The sequential approach was formally introduced by the government in the latest revision of PPG6 in 1996 in relation to retail and other town centre uses as a key consideration in decision-making. The approach is still relatively new and open to interpretation. However, it has already had a major effect in decisions on proposed retail developments, as shown later in Chapter 8. The sequential approach applies equally to both local planning authorities and developers. Local planning authorities are expected to adopt a sequential approach in allocating sites for new development in their development plans (as long as there is an identified need or capacity for further developments). Both local planning authorities and developers selecting sites for development should be able to demonstrate that all potential town centre options have been thoroughly assessed before less central sites are considered for development. 'If a developer is proposing an out-of-centre development, the onus will be on the developer to demonstrate that he has thoroughly assessed all potential town centre options' (PPG6, para. 1.9).

The approach implies a clear order of preference for central sites. 'Adopting a sequential approach means that first preference should be for town centre sites, where suitable sites or buildings suitable for conversion are available, followed by edge-of-centre sites, district and local centres and only then out-of-centre sites in locations that are accessible by a choice of means of transport' (PPG6, para. 1.11).

The guidance states that all planning applications for retail developments over 2,500 square metres gross floorspace should be supported by evidence on whether the applicant has adopted a sequential approach to site selection and the availability of suitable alternative sites. The guidance also makes important points about flexibility and the criteria for assessing the acceptability of potential sites.

> The government recognises that the approach requires flexibility and realism from local planning authorities, developers and retailers. Developers and retailers will need to be more flexible about the format, design and scale of the development, and the amount of car parking, tailoring these to fit the local circumstances. Local planning authorities should be sensitive to the needs of retailers and other town centre businesses and identify, in consultation with the private sector, sites that are suitable, viable for the proposed use and likely to become available within a reasonable period of time.
>
> (PPG6, para. 1.12)

The advice on flexibility implies that an alternative site may still be acceptable for the purposes of the sequential approach even if it cannot accommodate a development to the same scale or form as originally proposed (Holt, 1998). The requirement for sequential site assessment raises the issue of whether developers will adopt a 'built form' approach to new proposals for retail development or whether they will be required to use a 'class of goods' approach. The built form approach means that developers will look for sites that can accommodate the built form of development that they want, e.g. a superstore with adjacent and plentiful car parking. However, there is little possibility of finding sites for this type of development in town centres. The alternative, which is favoured by the government, is to use a class of goods approach which says 'can the class of goods which are proposed in a development scheme reasonably be traded in the town centre?' For instance, it could be argued that a proposal for the relocation of a town centre supermarket to an out-of-centre or even edge-of-centre site should not be allowed if it results in the loss of an existing town centre shopping facility. The 'class of goods' test would say that the facilities proposed are already available in the town centre and that PPG6 requires town centres to be sustained and enhanced. However the advice is interpreted, the problems of finding town centre and edge-of-centre sites could lead to continuing pressure from retailers to develop in out-of-centre locations.

Flexibility in format should not disregard the commercial realities of retailing. This point was recognised by an inspector in allowing a retail warehouse development on appeal in Birmingham in June 1999. The inspector said that it was not necessary for the appellants to 'abandon a retailing format that has proved successful and shoehorn a smaller development onto a smaller site where it is likely to replicate existing provision in the area'.

PPG6 expects developers to demonstrate rigorousness in the search for alternative sites. There must be a thorough examination of alternatives, which suggests that a systematic assessment is required. The requirement for flexibility in the application of the sequential approach is shown in a call-in inquiry decision by the Secretary of State in 1998 on proposals for an Asda superstore in Walsall. In assessing the availability of alternative sites, Asda rejected an alternative suggested by the council because it was too small to accommodate a store of the size proposed (8,400 square metres). The inspector judged that the alternative was large enough to accommodate a smaller superstore and that PPG6 defined a superstore as having a minimum of 2,500 square metres gross. He said there was no reason to believe that another operator would not wish to occupy the site, and therefore it was an available alternative. Sites must be genuinely available. The availability of alternative sites may be questionable if they have been on the market for a long time without any development interest.

PPG6 does not use the term 'sequential test' but in effect the application of the sequential approach is a 'test' which must be satisfied if a proposed development is to be acceptable. The approach is simple in principle but more difficult in practice. It must be applied with great care. A proposed approach to the practical application of the 'test' has been put forward by Halman (1998), as follows:

- define the primary catchment area of the proposal and identify sites in or on the edge of centres capable of serving broadly the same area
- identify sites to be appraised in terms of suitability, viability and availability for the form of development proposed
- assess sites on their ability to accommodate broadly the same form and amount of floorspace proposed by the developer.

The Planning Minister, Richard Caborn, said in his 'clarification' of PPG6 in February 1999:

> In applying the sequential approach, the relevant centres in which to search for sites will depend on the nature and scale of the proposed development and the catchment that the development seeks to serve. The scale of such proposals should also be appropriately related to the centre – whether town, district or local – the development seeks to serve.
>
> (DETR, 1999a)

Therefore, alternative sites should be in broadly the same catchment area, and it is not necessary to examine sites which serve a different catchment area, say in nearby towns. The minister also referred to extensions of existing stores. He said:

The sequential approach applies equally to proposals for extending existing edge-of-centre and out-of-centre development which creates additional floorspace. Local planning authorities should treat extensions as if they were new development.

(DETR, 1999a)

The interpretation of the sequential approach has become a matter of considerable contention (Holt, 1998). The most contentious element is the definition of 'edge-of-centre'. There is a wide range of interpretation of what is edge-of-centre according to the circumstances. The main test is 'walkability' or the ease of making linked trips to and from the town centre (Halman, 1998). Edge-of-centre is defined in PPG6 Annex A as: 'for shopping purposes, a location within easy walking distance (i.e. 200–300 metres) of the primary shopping area, often providing parking facilities that serve the centre as well as the store, thus enabling one trip to serve several purposes'.

The guidance also states (PPG6, para. 3.14):

Edge-of-centre locations will be determined by what is an easy walking distance for shoppers walking to, but more importantly away from, the store carrying shopping. The limits will be determined by local topography, including barriers to pedestrians, such as major roads and car parks, the strength of attraction of the town centre, and the attractiveness of the route to or from the town centre. However, most shoppers are unlikely to wish to walk more than 200 to 300 metres, especially when carrying shopping. The definition of edge-of-centre will vary between places, with large centres usually able to attract people to walk further than small centres.

The interpretation of what is 'edge-of-centre' can be seen in decisions on planning appeals and call-ins (Holt, 1998). Locations which are more than 300 metres from the primary shopping area can still be regarded as edge-of-centre if:

- there are no constraints to pedestrian movement
- the route is attractive or interesting
- there is good visibility between the site and the primary shopping area
- there are good physical linkages.

A superstore was allowed on appeal in Tetbury, Gloucestershire, in October 1999 even though it was more than 400 metres from the boundary of the town centre, on the grounds that there was an identifiable need for a new store and there were no possible alternative sites. The store would encourage linked trips.

On the other hand, locations which are less than 200 to 300 metres from the primary shopping area can be regarded as not edge-of-centre if there are barriers to safe pedestrian movement, such as major roads, if there are difficult gradients, and if there is not an easy route to follow. Acceptable walking distances will usually be less for small centres than for large ones.

The research report, 'The impact of large foodstores on market towns and district centres' (CB Hillier Parker, 1998), takes the view that the current distance guidelines of 200 to 300 metres for edge-of-centre locations in PPG6 may be too wide for some small market towns. It says:

> In addition to the need for strong physical links with the town centre, edge-of-centre development should be of an appropriate scale relative to the centre, and complement the existing retail offer. Local authorities need to help 'create' linkages to ensure edge-of-centre stores complement rather than supplant the convenience shopping role of these centres.
>
> (CB Hillier Parker, 1998, para. 61)

The sequential approach, therefore, requires a subjective but rigorous assessment of a range of factors which apply to alternative sites. The advice on best practice is to regard sequential site assessment as a 'test' which has to be satisfied to comply with the guidance in PPG6. With reference to the terminology in PPG6, the following factors should be examined for each alternative site:

- *suitability* – the current use of the site, the size of the site and its capacity to accommodate the type of development proposed; traffic and highway issues; the relationship between the development plan strategy and the proposals.
- *availability* – the ownership and ease of site assembly; the availability for development in a reasonable period of time (up to two to three years).
- *viability* – the financial viability of the development in relation to development costs and projected values.

Sustainability tests

The guidance in PPG6 on assessing new retail developments states (paragraph 4.1) that such developments should support the government's objectives of sustaining and enhancing existing centres and should be in accord with the strategy for retail development set out in the development plan. Where developments are proposed outside existing centres, a number of key tests will apply, as shown earlier in Figure 5.1. In addition to the quantitative 'impact test', there are three tests concerned essentially with sustainability:

- accessibility by a choice of means of transport
- impact on travel and car use
- (where appropriate) any significant environmental impacts.

These are discussed in turn in this section. The sequential test also aids sustainability because in edge-of-centre sites there is a strong possibility of linked trips and reasonable public transport or access by cycle or on foot (Holt, 1998).

Accessibility

The accessibility test means that developments should preferably be located in or next to town centres, in locations which are well served by public transport, or are easily accessible on foot and bicycle. Where new retail development is proposed away from town centres, the local planning authority should identify and appraise its likely accessibility by a choice of means of transport. On accessibility, PPG6 states:

> For new retail developments, local authorities should seek to:
>
> - establish whether public transport will be sufficiently frequent, reliable, convenient and come directly into or past the development from a wide catchment area;
> - ensure that the lack of public transport in rural areas should not preclude small-scale retail or service developments where this would serve local needs; and
> - ensure that the development is easily and safely accessible for pedestrians, cyclists and disabled people from the surrounding area (PPG6, para. 4.8).

Wherever there is a clearly defined need for major travel-generating uses which cannot be accommodated in or on the edge of existing centres, it may be appropriate to:

- combine them with existing out-of-centre developments
- negotiate for improvements to public transport accessibility to maximise access by means other than by car and to increase the ability for single trips to serve several purposes (PPG6, para. 1.17).

It is common for developers of edge-of-centre and out-of-centre schemes to offer improvements in public transport facilities to meet the accessibility requirements of PPG6. Measures may take the form of a shuttle bus service or a free bus service provided by the store operator. Developers are often obliged to assist with contributions to new or improved public transport, pedestrian and cycle facilities. Another factor is accessibility on foot in terms of the amount of residential walk-in population living near to a proposed development. Walk-time isochrones of 5 and 10 minutes are sometimes used to give an indication of the potential for walk-in trade, which represents an area of up to 750 metres from the development. However, a maximum distance of 500 metres is probably more reasonable (CB Hillier Parker, 1998).

The accessibility of town centre and edge-of-centre sites facilitates linked trips. The DETR/CB Hillier Parker research study on large foodstores in market towns comments on linked trips, as follows:

> The propensity to undertake linked trips depends on four inter-related factors:
>
> the extent to which the store complements the town centre/district centre;
>
> the distance and physical linkages between the two;
>
> the relative size of the centre as compared with the store; and
>
> accessibility, parking and orientation of the store.
>
> Shoppers using town centre foodstores are more likely to undertake linked trips in the centre than shoppers using an edge-of-centre foodstore. Similarly, shoppers using edge-of-centre foodstores are in general more likely to undertake linked trips with the centre than those using out-of-centre foodstores. This appears to support the policy preference for town centre or edge-of-centre stores.
>
> (CB Hillier Parker, 1998, para. 46–47)

The advice on best practice with regard to the 'accessibility test' is to:

- examine existing bus routes, their degree of penetration into the catchment area of the proposed development, and the frequency of services
- determine the accessibility of the location of the proposed development by bus and on foot
- identify the location of existing bus stops
- examine how bus services to the site could be improved to maximise the use of buses by shoppers and employees
- assess the potential for improved pedestrian linkages/pedestrian crossing facilities
- assess the potential for access by cyclists
- consider the extent to which the development would facilitate linked trips and the likelihood of shoppers making linked trips.

Travel and car use

Traffic and highway factors have always been important in assessing the impact of a new retail development. Proposals for development which are potentially large traffic generators will usually be required to be accompanied by a traffic impact assessment (TIA). Now that sustainable development is a prime objective of government planning policy, transport issues concerned with development proposals have become much broader. A traditional TIA will no longer satisfy the requirements of PPG6 and PPG13. 'Transport assessments' will replace the existing use of TIAs and the DETR is to issue good practice advice on their content and preparation. They are to be submitted alongside applications for major development proposals. In the case of shopping development the threshold is 1,000 square metres gross floorspace (DETR, 1999c).

> The government is seeking, through the location of development, to influence overall levels of car travel. PPG13 seeks to reduce the need to travel, reduce reliance on the car and facilitate multi-purpose trips. It therefore sets out policies for locating major generators of travel demand in locations which are, or are capable of being, well served by public transport.
>
> (PPG6, para. 4.9)

The key aim of the guidance in the existing PPG13 is to ensure that the local authorities carry out their land use policies and transport programmes, including their development control powers, in ways which help to:

* reduce growth in the length and number of motorised journeys
* encourage alternative means of travel which have less environmental impact, and
* reduce reliance on the private car.

In this way, PPG13 expects the local authorities to make a significant contribution to the government's sustainable development strategy.

'The guide to better practice' (1995), published to accompany PPG13, gives advice on the application of locational principles to different types of development including retail uses. It states that shopping accounts for more journeys (20 per cent) than any other purpose except journeys to work (22 per cent). Since the mid-1970s there has been a 27 per cent increase in the number of shopping trips, but a 57 per cent increase in the distance travelled (assisted by the growth of out-of-centre retailing). By the mid-1990s half of all shopping trips were undertaken by car.

Government policy aims to limit the amount of car-borne shopping. PPG6 states:

> For retail developments, local planning authorities should assess the likely proportion of customers who would arrive by car and the catchment area which the development seeks to serve. Particular

consideration should be given to retail proposals which seek to attract car-borne trade from a wide catchment area.

(DoE 1996, para. 4.11)

The achievement of sustainable travel objectives means not only a higher proportion of shopping trips by modes of transport other than the private car, but also a reduction in the length of car journeys on shopping trips. In relation to foodstores, the DETR/CB Hillier Parker study states:

> Four factors are likely to affect how vehicle travel distances may be altered by a new store. These are:
>
> the difference in distance travelled to the new store compared with the old;
>
> the change in mode;
>
> the change in frequency; and
>
> the change in propensity to undertake linked trips.
>
> The change in distance travelled will depend on individual circumstances, but in certain cases the distance travelled will decrease as people are 'clawed back' from a more distant store to a more local one. This will depend on a range of factors including the current availability of modern foodstores within the local area. Car use is likely to increase by a small degree as a result of a new store. Frequency of trip is unlikely to be a significant factor. Similarly, there is no evidence that the propensity to link trips will either increase or decrease.
>
> (CB Hillier Parker, 1998, paras 39–42)

The same report refers to significant debate over whether a detailed quantitative travel distance assessment for new stores should be undertaken. It says:

> Our experience suggests that such assessments are imprecise due to the complexity of travel patterns. Such patterns are dependent on people's lifestyles and can vary from day to day. In addition, any changes, either positive or negative, are likely to be very small in the context of the overall distance travelled for food shopping. These changes are unlikely to be a determining factor in the decision making process of where to locate a new foodstore; they are likely to be of more relevance in the case of other forms of retail and leisure development.
>
> (CB Hillier Parker, 1998: 104)

Nevertheless, transport assessments of proposed retail developments commonly look at the likely reduction in travel distance which would occur. The methodology is based on the following steps:

- Compare the number of trips to existing stores and the new store before and after its opening. The total number of trips to each store can be predicted using the trip rate information computer system (TRICS) database which gives trip rates as two-way totals per day per 100 square metres gross floor area for food superstores, discount foodstores, DIY stores and retail parks.
- Depending on the type of store proposed, the distribution of these trips by the sub-area or zone of origin can be estimated using the expenditure flow matrix which forms part of the 'best practice' approach to the assessment of quantitative impact.
- The number of trips from each sub-area to each store, before and after the opening of a new store, is multiplied by the distance in kilometres. The resulting totals of vehicle kilometres before and after opening are compared.
- The changes in length of car journeys can be expressed as percentage reduction in distance, reduction in the average length of trips, and total saving in vehicle kilometres per annum.

An important factor in the 'travel and car use test' is clawback of leakage of trade.

> Arguments may be advanced that more stores would lead to less overall travel or would prevent trade 'leaking' away to more distant centres. Local planning authorities should consider such arguments and also whether a more central location and/or another store in a district or local centre would:
>
> ensure easier access to all customers;
>
> facilitate more linked trips; and
>
> help achieve the overall aim of reducing reliance on the car for all trips.
>
> (PPG6, para. 4.10)

The DETR/CB Hillier Parker study examined the clawback argument. It said:

> Significant claims are made concerning the ability of new superstores to claw back trade to the centre in question, and the retail benefits derived from this. Our analysis suggests that the extent to which new foodstores will claw back trade will depend upon:

the size and accessibility of the store;

the location of the store (out-of-centre or edge-of-centre);

the nature of the catchment area (the range and proximity of competing foodstores); and

whether any other non-central foodstores have been developed in close proximity to the town.

Where there is already a well established non-central foodstore . . . it is unlikely that an additional edge-of-centre store will achieve the same level of clawback. Our research shows that large, highly accessible superstores are likely to achieve higher levels of clawback than smaller, less accessible stores, irrespective of location.
(CB Hillier Parker, 1998, paras. 43–45)

The advice on best practice with regard to the travel and car use test is to:

- analyse the characteristics of the catchment population in terms of car ownership, and household survey data on travel patterns for shopping
- assess the proportion of car-borne trade likely to be generated within the catchment area
- estimate to what extent the proposed development will claw back leakage of trade from more distant centres
- calculate the likely reduction in the length of car journeys by shoppers.

Environmental impacts

The Town and Country Planning (Assessment of Environmental Effects) Regulations were published in 1988 and revised in March 1999, accompanied by DETR Circular 2/99. The regulations are relevant to retail developments which may be Schedule 2 applications, that is proposed developments which would be likely to have significant effects on the environment by virtue of their nature, size and location. Schedule 2 has been amended. Infrastructure projects (category 10) now comprise urban development projects including the construction of shopping centres and car parks, sports stadia, leisure centres and multiplex cinemas. The criteria to be applied to such projects are whether the ground area which would be covered by the proposed development exceeds 0.5 hectares. The new regulations, like the old ones, should be applied in the context of the likelihood of significant environmental effects in terms of the size, nature and location of the proposed development. The criteria of site area have been reduced but the PPG6 floorspace thresholds still apply. Referring back to the 1988 regulations, PPG6 (para. 4.19) advises that the need for an EIA of major shopping proposals should be considered in the light of the sensitivity of the particular location. The thresholds are 20,000 square metres gross for out-of-town schemes, and 10,000 square metres gross for new retail proposals in urban areas on land that has not been previously intensively developed.

It is very unlikely that any retail developments other than major shopping centres or mixed use retail and leisure schemes would be judged to require an EIA. Potential environmental impacts associated with major new retail developments include:

- traffic congestion and air pollution
- the effects on ecology
- the visual impact on the landscape
- the effects on historic townscapes or conservation areas.

The detailed questions concerning these areas of potential impact lie beyond the scope of this book.

On the question of procedure, for all Schedule 2 developments the local planning authority must determine whether or not an EIA is required. Developers can obtain a formal (scoping) opinion from the local planning authority on what should be included in the environmental statement. In line with the sequential approach, a developer has to include in the environmental statement an outline of the main alternatives considered and the reasons for his or her choice. For those schemes where an EIA has to be prepared, detailed guidance is available on the procedures to be used. In particular, the DETR has published a good practice guide for the preparation of environmental statements for planning projects that require environmental assessment (DETR, 1995).

Interpretation of the significance of retail impact

The best practice advice so far in this book has covered questions of quantitative impact (the 'impact test') and the qualitative factors emphasised in PPG6. However, as shown in Figures 6.2 and 6.3, the final judgement on impact requires interpretation of the implications of a predicted level of impact in the context of the vitality and viability of centres. There is still great uncertainty about the significance of a particular level of impact on a particular centre. The DETR/CB Hillier Parker research study on foodstores (1998) noted that the calculation of predicted trade diversion is only the starting point for a proper assessment of the effects of proposed new stores. It states that in every case it is the implications and interpretation of the forecast levels of impact which determine the acceptability or otherwise of such proposals.

Because there is a direct relationship between quantitative impact and the vitality and viability of a centre, it follows that the significance of a particular level of impact on a particular centre can be judged by a 'model of

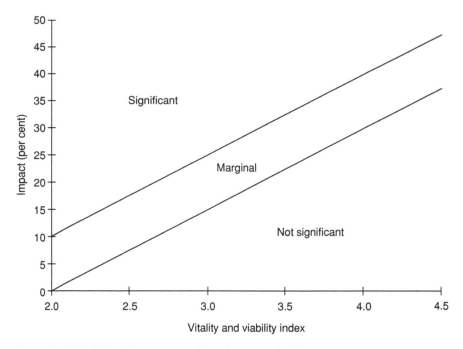

Figure 6.2 Model of significance of retail impact for foodstores

significance' of retail impact. The 'model' is a graphical representation of the relationship between percentage trade diversion and the vitality and viability index based on the town centre appraisal. It has been devised in the light of the experience of a wide range of retail impact studies and planning appeals where retail impact factors were influential in the final decision. The parameters used in the model of significance vary between foodstores and non-food retail developments because the levels of trade diversion are generally higher for foodstores. The model applicable to foodstores is shown in Figure 6.2.

If the percentage impact of a proposed superstore is high (say 20 per cent or more) the significance of that impact will be much greater for a centre with a below-average index of vitality and viability (3.0 or below) than for a centre with a high index of vitality and viability (say 4.0 or above). If percentage impact is low (say below 10 per cent) its significance will tend to be low for a centre with a vitality and viability index of above 3.0, but there may still be a significant impact if a centre has a low level of vitality and viability (say below 2.5). Similarly, if the vitality and viability of a centre is average (3.0) it will take a relatively large impact (over 25 per cent) to be significant. A smaller impact (less than 15 per cent) may not be significant. The model applicable to non-food retail developments is shown in Figure 6.3.

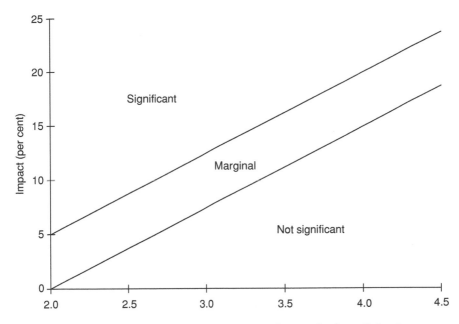

Figure 6.3 Model of significance of retail impact for non-food retail developments

In relation to non-food or comparison shopping developments, the model is equally applicable to retail warehouses and parks, factory outlet centres and other large out-of-centre developments which may still be 'in the pipe-line' as unimplemented planning consents. The model has been tested with data on all these types of development. It confirms the significance of the adverse impact of particular regional shopping centres (Merry Hill, Metro Centre and Meadowhall) and some large out-of-town retail parks which were refused consent on impact grounds (Teesside Park extension and the Centre 21 proposals, Leicester). It also confirms the lack of significance of impact of most smaller retail park schemes and factory outlet centres.

If the percentage impact of a proposed non-food retail development is high (say 10 per cent or more) the significance of that impact will be much greater for a centre with a below-average index of vitality and viability (3.0 or below) than for a centre with a high index of vitality and viability (say 4.0 or above). If percentage impact is low (say below 5 per cent) its significance will tend to be low for a centre with a vitality and viability index of above 3.0, but there may still be a significant impact if a centre has a low level of vitality and viability (say below 2.5). Similarly, if the vitality and viability of a centre is average (3.0) it will take a relatively large impact (over 12 per cent) to be significant. A smaller impact (less than 8 per cent) may not be significant.

The differences between the graphs in Figures 6.2 and 6.3 are in the scale of percentage impact on the y-axis. For foodstores, the levels of percentage trade diversion need to be twice as much as for non-food retail developments to have a significant impact. Foodstores operate at relatively high levels of turnover per square metre. They can often withstand high percentage impacts (more than 20 per cent) and still continue to have viable levels of residual turnover. Non-food retail developments usually form the major part of the shopping provision in town centres, especially in the larger city centres. A comparison goods trade diversion of more than 10 per cent could seriously harm a town centre's overall vitality and viability.

In both the foodstore and non-food retail development graphs there is a 'grey area' where the level of significance can be considered marginal. The significance of a predicted level of retail impact will depend on local circumstances. For instance, taking a proposed foodstore, there will always be a margin of error in the level of percentage impact predicted because the assessment of quantitative impact is inherently dependent on the assumptions used and the subjective allocation of trade diversion. The vitality and viability index should be more reliable but is still subjective in nature. Interpreting a marginal level of impact places even more emphasis on the qualitative factors outlined in this chapter.

The 'model' is a simple guide to interpreting the significance of a certain level of retail impact. It is not intended to be applied as a 'rule of thumb' in assessing impact in broad terms. Further research is needed to test the approach thoroughly but there is sufficient evidence from the preliminary testing that has been done to be confident that the model is capable of assessing the significance of retail impact. The approach needs to be refined using empirical data on different types of retail development to set the parameters of significance but it should be a useful tool in helping to improve the application of the 'impact test' in PPG6. It could be extremely helpful in practice to local authorities and inspectors as a first step in judging the predicted impact of a proposed development. It should be particularly relevant to interpreting the factors in paragraph 4.3 of PPG6. A 'significant' impact judged by the model in Figures 6.2 and 6.3 could be interpreted as harming the strategy for a town centre and future private sector investment, and have a negative effect on the quality, attractiveness and character of a centre, its physical condition, its range of services, and the number of vacant properties.

Summary

Retail impact must be assessed in qualitative as well as quantitative terms. PPG6 lists qualitative factors to be considered is assessing proposals which may have an impact on the vitality and viability of centres. Vitality and viability has become the cornerstone of government policy on sustaining and enhancing town centres. PPG6 emphasises the need to assess vitality and viability using health checks. In this chapter the PPG6 health check indicators and corresponding URBED factors have been combined to produce an integrated checklist of factors that are relevant to the appraisal of a town centre. The appraisal framework is based on a five-point scale. It is a subjective but systematic approach to assessing vitality and viability and it should be applied in all RIAs.

Qualitative need includes issues of the nature and quality of existing shopping provision, indications of over-trading, and consumer demand. It is necessary to apply subjective judgement, based on the town centre appraisal, to assess if there are deficiencies in the type and quality of existing shopping provision, and to consider the wider needs of the community.

PPG6 places emphasis on the sequential approach to selecting sites for retail development. Local planning authorities must adopt a sequential approach in allocating sites for new development. If proposals are made for an out-of-centre development, a developer must demonstrate that all potential town centre sites have been thoroughly assessed. The sequential approach means that the first preference is for town centre sites, followed by edge-of-centre sites, district centres and local centres, and only then out-of-centre sites which are accessible by a choice of means of transport. The approach requires flexibility and realism from local planning authorities, developers and retailers in the format, design and scale of development. It requires a rigorous assessment of alternative sites to assess their suitability, availability and viability for development.

The 'accessibility test' in PPG6 means that developments should preferably be located in or on the edge of town centres, in locations which are well served by public transport, or are easily accessible on foot and by bicycle. The accessibility of town centre and edge-of-centre sites facilitates linked trips. In applying the accessibility test, it is necessary to examine the adequacy of existing bus services, ease of access for pedestrians and cyclists, and the likelihood of shoppers making linked trips.

Government policy aims to reduce the need to travel and reduce the reliance on the car, including seeking to limit the amount of car-borne shopping. The achievement of sustainable travel objectives implies a higher proportion of shopping trips by public transport and a reduction in the length of car journeys on shopping trips. An important factor in the 'travel and car use test' is clawback of leakage of trade. A new store could reduce travel distance by clawing back leakage.

The EIA regulations are relevant to proposed new retail developments which may have significant effects on the environment because of their nature, size and location. It is very unlikely that proposals for retail development other than major shopping centres or mixed use retail and leisure schemes would be judged to require an EIA.

The implications of a predicted level of retail impact need to be interpreted in the context of the vitality and viability of centres. A 'model of

significance' of retail impact has been devised to show the relationship between predicted trade diversion and the vitality and viability index of a
centre based on the town centre appraisal. The model should be applied to
foodstores and non-food retail developments separately, and is intended as a
simple guide to interpreting the significance of a certain level of retail
impact. It is a useful tool in helping to improve the application of the
'impact test' in PPG6.

7

EVIDENCE OF RETAIL IMPACT

The experience of the impact of new types of shopping facilities in Britain varies between different types of development – foodstores, retail warehouses and parks, factory outlet centres, and regional shopping centres. In this chapter reviews are made of each type of development and the evidence of their impact. It concentrates on the lessons to be learnt from the available evidence. Most of the concern about retail impact was initially related to superstores and most RIAs are still carried out on proposed superstore developments. However, there can be impact issues in all types of retail development. PPG6 refers to impact issues arising from the development of discount foodstores, retail parks, regional shopping centres and factory outlet centres. It states that the government wishes to provide a common framework for handling applications for different types of retail development to ensure that retailers have confidence that applications will be handled on a consistent basis.

Foodstores

Superstores

The debate about the impact of new shopping developments and their planning implications can be traced back to the earliest origins of superstores in Britain. A superstore is defined as a 'single-level, self-service store, selling mainly food, usually with more than 2,500 square metres of trading floorspace, with supporting car parking'. The main operators usually seek sites for stores of between 4,000 and 5,000 square metres of selling space. From an innovation phase in the 1960s, there was a period of resistance to the development of superstores in the 1970s, followed by a phase of renewed development in the early 1980s, reflecting the attitude of public policy towards superstore development (Davies and Sparks, 1989).

The development of superstores in Britain dates from the abolition of resale price maintenance in 1964 which led to economies of scale in building large grocery stores seeking sites in suburban or edge-of-town locations. All the major grocery retailers embarked on programmes of large store development in the 1970s and there was intense competition between operators for sites (Guy, 1988). The growth of superstores has been quite dramatic. In the 1980s the rate of expansion in the number of superstores was more rapid than the growth in the volume of food sales. Therefore, the rise of superstores was not supported by expenditure growth and resulted in a loss of market share amongst small independent food retailers, e.g. butchers and smaller supermarkets. Between 1982 and 1992 the number of large grocery outlets and other food retailers declined by a similar amount in relative terms (about 30 per cent) but the absolute decline in numbers of smaller food retailers has been considerable. The total turnover of all food retailers increased by 100 per cent between 1982 and 1992. The increase in turnover of large grocery retailers was 159 per cent whereas the turnover of other food retailers increased by only 7 per cent.

The extraordinary performance of Britain's leading grocery retailers during the 1980s and early 1990s was an era of 'store wars' between the major grocery retailers in the new store development process. There was intense competition between the major companies for the most attractive development sites (Wrigley, 1991). Superstore operators were prepared to pay up to £5 million per hectare for the best sites and the major operators all pumped huge investments into new store expansion (Wrigley, 1992). There was a growing concentration of market share among the leading operators. The pressure for superstore development has not diminished in spite of a tightening of government policy in the 1990s. The number of superstores in Britain increased from less than 500 in 1986 to about 1,100 in 1999. Over the same period superstores increased their share of total grocery sales from 30 per cent to 54 per cent (CB Hillier Parker, 1998). Superstores are expected to take an increasing share of convenience shopping in the future. The Institute of Grocery Distribution has predicted that the combined market share of the 'big four' operators (Tesco, Sainsbury's, Asda and Safeway) will increase from 45 per cent in 1998 to 53 per cent in 2005.

Some commentators believe that the grocery market in Britain is becoming 'saturated' and that this a cause for concern. The growth in the number of superstores is slowing down and saturation point could be in the order of 1,300 stores. Guy (1996) notes that 'saturation' can be defined in two ways: a situation in which the local population is well served by a choice of modern grocery stores, such that there is no widespread desire for more to be provided, and a situation in which any major new store can only be viable if its entry is matched by the demise of one or more existing grocery stores.

Guy (1994c) comments that saturation is a simplistic concept which is meaningless when used at the national level. It can only be applied at the local level. Some parts of Britain are considered to be saturated but in others there are still opportunities for new development. He thought that in the 1990s, rather than increasing their overall share of the market, the leading grocery companies would take market share from each other at the local level.

There is considerable pressure for further expansion in the retail grocery market and the main operators are looking to maintain or increase their market share, e.g. through developments in regions where they have not traditionally been strong. Langston et al. (1997) consider that there is considerable potential for further expansion in the British grocery market, but the new planning guidelines in PPG6 may in the long run affect the type of foodstore development that takes place by limiting out-of-centre growth and focusing attention on edge-of-centre sites. As government policy restricts the availability of out-of-centre sites, a shift in emphasis is taking place towards smaller food stores rather than superstores, such as Sainsbury's country stores and Tesco Metro which are designed for small towns and suburban centres to provide essentially for lunchtime and basket top-up shopping. Another current trend is towards superstores selling an increasing proportion of comparison goods, e.g. clothes and footwear, or incorporating in-store pharmacies or post offices. There can be an adverse impact from large food stores on existing shops as their product range is broadened to include non-food goods. Some operators, e.g. Asda, are seeking to build larger stores to accommodate an increasing amount of non-food floorspace.

Discount foodstores

The recession of the late 1980s and early 1990s, coupled with the perceived move 'up-market' by the major food retailers, left a niche to be exploited by the food discount operator Kwik Save and the newer continental discounters such as Aldi, Netto and Lidl. By 1993 food discounters controlled around 10 per cent of the overall grocery trade in the UK and it was thought that their share of the grocery market could reach 14 per cent by the end of the decade (Arnold, 1995). However, there is evidence that their market share in Britain is not growing as quickly as was originally expected. Aldi, Lidl and Netto still account for only a small proportion of the UK grocery market (Eade, 1999). In other European countries, discounters are more established and have a higher market share, e.g. 20 per cent in Germany.

Discounters have low profit margins (typically 4 per cent or less) and offer a 'no frills' approach. They operate from cheap sites, have stores with simple fittings and shop fronts, few staff, a limited product range and a high turnover of goods. Prices are usually at least 10 per cent lower than in the larger foodstores and they tend to cater for the less affluent, often non car-borne shopper (Grimley, 1993). Discount foodstores are generally less than 2,000 square metres gross and occupy sites of no more than a hectare. The net floorspace is usually only 700 to 800 square metres. They sell between 500 and 3,000 lines, depending on the operator, compared with about 20,000 to 25,000 lines at a typical superstore. The four main discount operators (Kwik

Save, Aldi, Netto and Lidl) each have defined site criteria which ensure that sites are the right shape and large enough to accommodate the preferred store size and layout.

Discounters tend to pose little threat to the leading superstore operators who are increasingly marketing quality and customer service rather than price, although the major superstores have responded by making some price reductions and stocking cut-price basic items. There are signs of a complementary relationship between superstores and discount stores, and discounters are often keen to locate close to existing out-of-centre superstores. The discount sector is growing rapidly and, together with the continued success of superstores, there appears to be a trend towards polarisation of the grocery market (Hogarth-Scott and Rice, 1994). Local authorities have generally not been opposed to the discount store boom. Since 1990, planning permission has been granted in at least 75 per cent of planning applications for discount stores and there appears to have been no decline in the rate of approvals since the July 1993 version of PPG6. Of those applications which have been refused, few are for reasons of retail impact (Arnold, 1995).

Evidence of the impact of foodstores

The first superstore inquiry where impact assessments were presented was the Carrefour hypermarket at Chandlers Ford, Eastleigh, in 1971 (Noel, 1989). The economic impact of superstores and hypermarkets was most extensively studied in the 1970s but shopping issues arising out of applications for the development of superstores have remained important at inquiries up to the present day. It has been estimated that 'shopping issues' of this kind accounted for just under 40 per cent of refusals of planning permission for grocery superstore development between 1971 and 1991 (Lee Donaldson Associates, 1991). The issue of the cumulative impact of a number of proposals has become of more recent interest. The BDP/OXIRM report (1992) stated that little work had been done on the cumulative effects of superstore openings. Early studies of the impact of superstores and hypermarkets on patterns of shopping behaviour focused on the trading characteristics of individual stores. Despite initial fears about the adverse impact of these developments on town centres, the evidence suggested that their effects were felt most by the smaller district centres in their immediate vicinity rather than by town centres (Bromley and Thomas, 1993).

The impact of superstores in the 1970s turned out to be much less severe than was first feared. Schiller commented that:

> The evidence is clear and consistent that the impact of freestanding superstores is diffused over a wide area. They have tended to divert a small amount of sales from a large number of surrounding centres. Superstores also have a relatively minor impact on small local shops.

In fact, the evidence is that the greater the size of competing stores, the greater the impact from the superstores.

(Schiller, 1981: 38)

However, Schiller notes that it is very difficult to distinguish the effects of superstores from other factors causing changes in spending patterns. Writing slightly later, Davies (1984) stated that most studies agreed that the main casualties of superstores and hypermarkets were the smaller branches of the multiple food retailers themselves. He concluded that 'overall, much of the concern that was expressed during the early years of the hypermarket debate has proved to be unfounded' (Davies, 1984: 276).

More recent evidence continues to support this view. Initial fears about the potentially drastic effects of superstores on older shopping facilities have gradually declined and superstore trading has emerged as the major feature of grocery shopping in Britain (Bromley and Thomas, 1993). During the 1980s there was a rapid increase in the number of superstores, a rapid rise in their total market share of food and convenience goods sales, and a marked shift to out-of-centre locations. There is increasing evidence, however, that the pace of this development is slowing, as noted earlier. The key issue about superstore development is how it will affect small market towns. The House of Commons Select Committee on the Environment report, 'Shopping centres and their future', reflected the concern that pressure for superstores in small towns could undermine the vitality and viability of these centres because it could reduce linked shopping trips for food and non-food goods (House of Commons, 1994). The market towns are now widely seen as a new growth area for the major food retailers (Parker, 1995). Further comments on impacts on market towns are made at the end of this section.

In the early years of superstore development, trade diversions of up to 10 per cent were generally thought to be acceptable. For example, an appeal into the refusal of planning permission for a 5,000 square metre Tesco superstore in Hartlepool was allowed in 1987. An impact of 9.5 per cent on town centre convenience trade was agreed between the parties and the inspector accepted that there was no evidence that the store would undermine the vitality and viability of the town centre as a whole. Later the threshold of acceptance by inspectors tended to rise to 15 per cent. In a call-in decision in 1995 relating to a 5,180 square metre gross food superstore in Trowbridge, Wiltshire, a 14 per cent impact was regarded by the inspector as 'on the margin of acceptability'. The proposal was refused by the Secretary of State on policy grounds because it was an out-of-centre location and the sequential approach had not been applied. An appeal was dismissed by an inspector in 1994 for a Safeway superstore of 5,850 square metres gross in Norwich on the grounds that impacts of 20 per cent or more were likely in two nearby district centres, which would undermine the vitality and viability of these centres.

A very high predicted level of impact will invariably lead to a refusal of an appeal or call-in. In 1999 the Secretary of State accepted an inspector's recommendation to refuse an application by Morrisons for a large superstore in Billingham on Teesside. The site was close to the town centre but separated from it by a dual carriageway road and by an area which would be an unattractive route for pedestrians. It was judged to be out-of-centre. The main issue was impact on the vitality and viability of Billingham town centre. Trade diversion was estimated by the parties to be between 54 and 66 per cent of town centre convenience turnover, depending on the estimated turnover of the Asda store, the main supermarket in the centre. The inspector concluded that trade diversion of this scale would be harmful to the town centre and to prospects for future investment. He considered that the existing Asda store, which underpins much of the current vitality and viability of the centre, would close.

Inspectors have sometimes judged that in the case of proposed foodstore developments a quantitative impact of less than 10 per cent on convenience trade may be unacceptable in the context of a centre operating at the margins of vitality and viability. In some cases, surprisingly low levels of impact can persuade an inspector to reject a superstore proposal. In a landmark decision on a call-in inquiry in 1995, an application by Morrisons for a large out-of-centre superstore in Hull was refused by the Secretary of State in agreement with the inspector's conclusions that harm to the continued prosperity of existing centres would be considerable. There was some dispute about the predicted trade diversions but the inspector took the view that the impacts on a total of six nearby centres would be in the range of 6.8 per cent to 10.8 per cent. Even at these modest levels of trade diversion there were serious concerns about the vitality and viability of existing centres.

In the 1990s superstore inquiries frequently had to deal with issues of cumulative impact where more than one proposal had to be considered. The importance attached to cumulative impact in the several versions of PPG6 has in some cases meant that a decision to allow an appeal for a superstore development in one inquiry becomes a material factor in deciding a later proposal. But in other cases more than one proposal has been considered at the same inquiry. Cumulative impact issues may dictate that there is only a potential for one development and that if more than one went ahead there would be an adverse impact on existing centres. It is then the responsibility of the inspector to decide on the planning merits of each proposal. An inspector can allow multiple proposals and let market forces determine which one goes ahead.

A good example of cumulative impact issues is the Secretary of State's decision in 1995 on a joint inquiry into four supermarket proposals in Cockermouth, Cumbria. One involved the change of use of the auction mart premises in the town centre and the other three were out-of-centre sites. The inspector concluded that there was a need for a new supermarket in Cockermouth, but there was capacity for only one out-of-centre store if the vitality and viability of the town centre was not to be harmed. All the parties agreed that a new store on any of the sites would have a convenience trade impact on the town centre of not more than 15 per cent. On balance the inspector decided that a new supermarket should be located in the town centre and recommended that the appeal in respect of the mart site should be allowed and the other three schemes rejected. This was accepted by the Secretary of State and planning permission was granted for a supermarket on the mart site.

High levels of cumulative impact can lead to proposals being unacceptable. An appeal by Tesco on proposals for an edge-of-centre foodstore of 1,724 square metres in Mablethorpe, Lincolnshire, was dismissed in 1999. The inspector concluded that the proposed development, in combination with an out-of-centre Lidl discount store which was just about to open, would have a cumulative trade diversion of up to 50 per cent of convenience goods turnover from Mablethorpe town centre. This would result in closures of shops in the town centre and undermine its vitality and viability. The appeal was dismissed.

The question of impact of discount food stores is a difficult one. They are individually small developments but their growing market share and ability to compete with supermarkets and other foodstores can raise issues of trading impact in certain situations. PPG6 recognises that discount food stores can sometimes have a significant impact on town centre retailing and it identifies the need to assess the likely impact of proposals for such developments on the vitality and viability of shopping in the town centre. The discounters are principally targetting the independents, the smaller Co-ops and Somerfield serving the lower social profile shoppers. Small towns and centres whose convenience trade is focused on independent retailers and smaller supermarkets serving the less mobile or affluent sectors are therefore disproportionately vulnerable to the discounter boom (Arnold, 1995).

Evidence of the impact of discount foodstores is available from a survey by the Scottish Grocers Federation in 1994. A survey was carried out of small independent retailers located near larger, cheaper discount stores. Twenty discount stores were chosen at various locations throughout Scotland and at each location ten independent retailers were surveyed within a 3-mile radius. Many retailers reported that they had lost a significant amount of business, an average of 11 per cent. Impact was particularly serious within a 1.5 mile radius, with turnovers down by over 12 per cent (Scottish Grocers Federation, 1994).

The evidence of appeals concerning discount foodstores also shows the importance of retail impact issues in influencing planning decisions. For example, an appeal against non-determination of proposals for an Aldi discount store in Prestatyn, North Wales, was dismissed in 1995. The main issue was the effect of the store, judged to be out-of-centre, on the vitality and viability of the retailing function of the town centre. It was agreed that the convenience goods trade diversion would be in the order of 8 to 10 per cent, which the inspector regarded as quite significant in a small centre which had already lost trade to a nearby Sainsbury's store. The reduction in the number of shoppers drawn into the town centre would have had a substantial impact on its vitality and viability.

Cumulative impact is also an issue in determining appeals concerning discount foodstores. An appeal was made by Aldi against the refusal of a proposed out-of-centre development in Boston, Lincolnshire, which was decided in 1994. Boston is a small town but it already had several large foodstores – Asda, Tesco, Somerfield, Co-op and a proposed Kwik Save. Aldi did not claim a quantitative need for further food shopping in Boston, but the store would provide a choice of quality and price not currently available in the area. The council acknowledged that the overall impact of an Aldi store on the town centre would be relatively small (6 to 8 per cent) but there was concern about the cumulative impact and the risk that the sole remaining town centre supermarket (the Co-op) would close. The inspector accepted that the vitality and viability of Boston town centre was fragile and concluded that although the individual impact of the proposed Aldi development would not be great, there was clear evidence that the cumulative effect would be to undermine the vitality and viability of the town centre. The appeal was therefore dismissed.

The research by CB Hillier Parker on large foodstores in market towns and district centres shows that large foodstores can and have had an adverse impact on these centres.

> The level, and consequences, of impact will vary depending on the particular local circumstances of the centres concerned. Smaller centres which are dependent to a large extent on convenience retailing to underpin their function, are most vulnerable to the effects of larger foodstores in edge-of-centre and out-of-centre locations.
>
> (CB Hillier Parker, 1998: 12)

The research identified impacts on the market share of the main food retailers in market towns and district centres of between 13 and 50 per cent as a result of large foodstores. This wide range of impact reflects a number of key factors:

the relative quality/attractiveness of the new store/operator;

location of the new store relative to the market town/district centre;

nature of the catchment area (i.e. whether there is already a dense network of competing foodstores); and the relative size of the new foodstore compared with the overall convenience floorspace in the centre itself.

(CB Hillier Parker, 1998: 112)

These levels of impact have led, directly and indirectly, to the closure of some town centre food retailers, increases in vacancy levels and a general decline in the quality of the environment of the centres concerned. Large falls in turnover do not necessarily mean that a store will close, but the consequences of impact are not necessarily immediately apparent. Turnover and profitability can decline over several years. The implications of a fall in turnover can also take time to manifest themselves in an identifiable change in the vitality and viability of a centre. This time lag can be several years. The final message of the study is that decision-makers should adopt a cautious approach to considering the location and likely long-term consequences of the development of large foodstores in non-central locations, and that there is a pressing need for improvement in the data and application of RIAs.

Retail warehouses and parks

Retail warehouses

Retail warehouses are defined in PPG6 as 'large single-level stores specialising in the sale of household goods (such as carpets, furniture and electrical goods) and bulky DIY items, catering mainly for car-borne customers and often in out-of-centre locations'. They are normally associated with the sale of 'bulky goods' in three main sectors: DIY/hardware, furniture and carpets, and electrical goods.

In planning terms the type of goods sold is of crucial importance since planning policy usually seeks to restrict the sale of goods which might be termed town centre comparison goods from out-of-centre locations. Store sizes vary according to the goods being sold but are generally 1,000 to 1,500 square metres in electrical goods stores; 3,000 to 4,000 square metres in DIY goods stores; and 4,000 to 5,000 square metres for furniture and carpets. Some of the leading retail warehouse operators, such as B&Q, Homebase and Toys R Us, have become so successful in terms of market share that they have been described as 'category killers'.

Retail warehouses first appeared in Britain in the late 1970s in response to the need for large-scale retailing of non-food bulky goods. There has been an evolution from early generation schemes involving conversion and reuse of industrial property through to the construction of purpose-built 'sheds' and then the development of large retail parks.

> In addition to the social trends which encouraged the general de-centralisation of retailing, other trends encouraging the development

of retail warehouses, particularly in the DIY and self-assembly furniture sectors, have been: the growth in home ownership, the increase in leisure time, and the rising labour costs of home improvement. The development of retail warehouses has been further encouraged by trends in expenditure – a rapid increase in demand for durable goods, particularly electrical goods, in contrast to virtually static expenditure on convenience goods.

(Gibbs, 1987: 19)

The long-term expenditure growth trends for bulky goods according to The Data Consultancy's Information Brief 98/3 are, in real terms: electrical goods – 7 per cent per annum, furniture and carpets – 2 per cent per annum, and DIY and hardware – 3 per cent per annum.

DIY retail warehouses and garden centres have always been viewed by planners as retail functions unsuited to trading within the fabric of traditional shopping centres because of their large space and car parking requirements. This view of competition as 'benign' fostered the concept of the complementary relationship between the retail warehouse phenomenon and the pre-existing retail system. In the context of the increasingly 'free market' attitudes of central government to retail planning in the 1980s, this view was maintained despite the incremental addition of increasingly specialised retail functions to this form of trading. This was accompanied by the emergence of the retail warehouse park in a variety of forms and scales.

(Bromley and Thomas, 1993: 134)

The range of goods sold from retail warehouses can be controlled through planning conditions or a Section 106 agreement which specifies what may or may not be sold. There is a move towards a 'negative' rather than 'positive' approach, specifying what *cannot* be sold. This might include goods which are essential to the health of a town centre, e.g. food, clothing and footwear, fashion accessories, jewellery, perfume, toiletries, books, music cassettes and CDs, etc. This type of approach can be more effective and more enforceable than a positive control on goods that can be sold. It avoids the need for a floorspace limit on certain uses and it can be defined flexibly depending on local circumstances. It is quite common for owners and operators of retail warehouses to seek to relax these restrictions to improve the trading performance of their stores.

Retail parks

Retail parks have experienced rapid growth since the mid-1980s. In 1999 there were almost 500 retail parks in Britain with an average of about

10,000 square metres of floorspace. PPG6 defines a 'retail park' as an agglomeration of at least three retail warehouses. There are also 'hybrid' forms of retail park, which include superstores or out-of-town comparison goods outlets such as Marks and Spencer. Notable examples are Fosse Park outside Leicester and The Fort Retail Park in Birmingham which is a hybrid between an out-of-town retail park and a small regional shopping centre. The distinction between retail warehousing and conventional retailing is becoming more blurred with new, traditionally town centre retailers moving into a sector previously limited to those selling bulky goods.

The main planning issues concerning retail parks are where they should be located and the extent to which they compete with town centres. Historically retail warehousing has been developed on out-of-centre sites. Free-standing retail warehouses were built mostly outside town centres, often on radial roads for good access by car. Retail parks were developed mostly where large sites could be found with adequate space for car parking, either on greenfield sites or as redevelopment schemes on brownfield land. Planning policy in the 1980s generally favoured out-of-centre retail warehouse development.

PPG6 recognises that large stores selling bulky goods may not be able to find suitable sites in or on the edge of town centres. Retail parks are often located out-of-town and that is not a major problem if it is accepted that they do not compete with town centres. But, as acknowledged in PPG6, the impact of retail parks depends on the range of comparison shopping that they offer. There are pressures from the retail industry for a relaxation on the sale of comparison goods outside town centres. The House of Commons Select Committee on the Environment (1994) identified the key to limiting the impact of retail parks as being the control of goods sold to prevent town centre comparison goods, e.g. clothes being sold out-of-town. But even where there have been restrictions to bulky goods there have been cases of 'planning creep' or trading-up over a period of time to more general retailing as defined in Class A1 of the 'Use classes order'.

The policy on retail warehouse parks was clarified by the Planning Minister, Richard Caborn, in a speech in November 1998. He said, with reference to retail warehouse parks:

> The main argument given for them has been for selling bulky goods that need to be taken away by car. In practice, many such retail warehouses either do not sell 'bulky goods' or, if they do, they are not taken away by car. Few people drive away with furniture or white goods. Firms may prefer larger showrooms, but in practice they could sell from town centre or edge-of-centre locations. We have challenged the need for the retail warehouse format. We have made clear that the sequential approach applies to the constituent parts, not whether a more central site can be found for the whole

development. We expect such developments to be more flexible about scale, format, design and car parking. We expect them to fit into existing centres.

(DETR, 1998)

This statement is confirmation that, although out-of-centre retail warehouses catering for car-borne trade are contrary to government policy, there is a clear preference for the sale of bulky goods in or on the edge of town centres. Edge-of-centre sites for retail warehouse parks, therefore, are consistent with government policy. Studies by England & Lyle show that edge-of-centre sites have a number of benefits as locations for retail parks:

- Edge-of-centre sites maximise the potential for spin-off benefits for their adjacent shopping centres by enabling shoppers to make linked trips to both the retail park and the town centre. Edge-of-centre developments can make a positive contribution to the vitality and viability of a centre.
- It is not necessary for a retail park to be within 200 to 300 metres of a town centre to be regarded as 'edge-of-centre' as long as there is an opportunity for linked trips to be made, and the site is accessible by public transport.
- Edge-of-centre developments minimise any possible harm to town centres through trade diversion, which is a drawback of some larger out-of-centre retail parks, and can enhance an adjacent centre.
- Edge-of-centre developments are compatible with PPG6 guidance, and in particular the sequential approach.
- Edge-of-centre locations are more sustainable than out-of-centre sites. They are generally more accessible by public transport as well as by car.
- Edge-of-centre sites offer other benefits. They are usually on brownfield sites, often on vacant or disused land where reuse for industry is not always viable. Therefore, edge-of-centre retail parks can make a significant contribution to urban regeneration.

Operators prefer to locate adjacent to similar users to create a critical mass to attract custom. Critical mass helps to create competition between retailers and it also benefits customers who then have a choice of outlets to visit. It also has PPG13 benefits in avoiding sporadic patterns of development. A large retail park can be a major shopping destination, similar to a town centre, where shoppers can make shared trips to purchase different items. However, accommodating large 'critical mass' parks adjacent to existing centres is difficult owing to the lack of large available sites, particularly adjacent to strong trading centres.

Retail parks are changing, with a polarisation of unit sizes and a more selective approach to location. Some are providing an increased range of products in much larger units, e.g. B&Q's warehouse stores. Others have decreased their floorspace and refined their range of products. Some companies are targeting stores in towns which traditionally would have been considered too small. Although in the 1980s the bulk of expansion focused on DIY and furniture, recent growth has been in white goods, e.g. toys and car parts, and in electrical goods and traditional high street goods such as sports and optical equipment. Leisure is also becoming a common feature of retail parks with, for example, multiplex cinemas and bowling alleys becoming increasingly popular.

Evidence of the impact of retail parks

The initial development of retail warehouses was considered unlikely to have an adverse effect on existing trading patterns. Generally there was less concern about the impact of retail warehousing because of the relatively low turnover per square metre which these stores tend to generate and the relatively high sales growth in this sector which tends to offset any trade diversion (Davies and Howard, 1988). However, the BDP/OXIRM report (1992) considers that the evolution of retail parks may offer a greater potential threat to traditional town centres than does out-of-town grocery retailing. There is now a much greater overlap with conventional town centre trading profiles, e.g. in electrical goods, furniture and toys.

There is little evidence of adverse trading impact of retail warehouses on established shopping centres. The evidence suggests that:

> Retail warehouses have particular appeal for car-borne shoppers drawn from a wide trade area. Thus their impact is diffuse and is likely to affect the town centres and district centres within their spheres of influence. The early notion of their essentially complementary relationship to traditional centres and their benign impact, however, appears in need of re-evaluation. The scale of decentralisation of the variety of new retail forms associated with retail warehousing suggests the existence of a more potent force for change than was initially anticipated. Retail warehouses are clearly capable of diverting a significant proportion of trade from existing centres for an increasingly specialised range of products.
>
> (Bromley and Thomas, 1993: 140)

Concern about the expansion of out-of-town retail parks into selling 'high street' durable shopping goods has also been expressed by the House of Commons Select Committee on the Environment. Its report comments that 'while individually the impact of such developments on the town centre may not be great, cumulatively they represent a relatively rapid shift of retail turnover from town centres to other locations (House of Commons, 1994: xxiii).

Inspectors at planning appeals for out-of-centre retail warehouses and parks have observed that it is often difficult to predict the impact on town centres. It is necessary to make a qualitative assessment of the performance of the town centre. The question of need may also arise, e.g. when there is leakage of trade out of an area. The degree of competition is important depending on the type of goods proposed. As with superstores, competing schemes may both be allowed by an inspector, leaving the market to determine which one would be built. For example, an inquiry was held in 1995 into proposals for two retail parks on out-of-centre sites in Dumfries. The reporter thought the estimated 12 per cent cumulative impact of the two schemes would be seriously detrimental to the town centre, but there were no clear-cut objections to either site. He allowed both on the proposition that the market would only allow one store to be built.

Impact considerations are not usually as significant for individual retail warehouses as they are for superstores because their turnover is much lower and they are part of a sector in which there is a relatively high growth of expenditure. Impacts are often seen as qualitative rather than quantitative. For instance, an appeal was decided in 1995 on an application for retail warehousing in Bath. The proposal would have drawn only 2 per cent of comparison goods expenditure in the area but the appeal was dismissed because the development was contrary to local planning policy on the location of retail warehouses in Bath. An appeal on a single retail warehouse proposal was also dismissed in Carlisle in 1995 because, although the inspector considered that in itself the store would have only a negligible effect on city centre turnover, there would be cumulative impact.

There are many other cases of appeals that have been allowed because impacts were considered to be insignificant. For instance, an appeal was allowed in Bolton in 1997 for an edge-of-centre retail park development. Retail impact was not a reason for refusal of the application; it was on land allocated for industrial use. It was accepted by all the parties that there would be no adverse impact on any established shopping centres. The inspector said that the development would positively enhance Bolton town centre and make a valuable contribution to the physical regeneration of a derelict site. The scheme, named Bolton Gate Retail Park, opened in 1998.

There are numerous examples of edge-of-centre retail parks which have been developed without any adverse impact. One is St Marks Retail Park in Lincoln which opened in 1994. It forms part of a major redevelopment scheme on the edge of Lincoln city centre, the newer part of which has been developed for town centre comparison goods retailing. The two developments are complementary. The retail park has met a need for bulky goods outlets which could not be accommodated within the city centre. There is no evidence that it has had any impact on the city centre, which is trading successfully, and it has generated additional trade through linked trips.

Where impact issues become more significant is in the case of very large retail parks, some of which are large enough to compare in size with existing town centres. In 1993 proposals were made for Phase 2 of Teesside Retail Park, comprising almost 10,000 square metres of non-food shopping in seven retail warehouse units. The application was called-in for determination by the Secretary of State. Evidence was presented to the inquiry that the proposed extension to the retail park would compound the serious adverse impacts which had already resulted from Phase 1, particularly on Stockton-on-Tees town centre. It was agreed that the combined impacts of Phases 1 and 2 on Stockton were likely to be between 15 and 20 per cent on comparison trade. The Secretary of State accepted the inspector's conclusions that the benefits of the proposals did not outweigh the harm which was likely to be caused to the currently fragile state of the town centre and to the policies which have been designed to protect its vitality and viability.

Factory outlet centres

PPG6 defines factory outlet centres as groups of shops, usually away from the town centre, specialising in selling seconds and end-of-line goods at discounted prices. They are outlets from which manufacturers, rather than retailers, can sell directly to the public, usually of 10 to 15,000 square metres gross floorspace containing 50 or more shop units selling end-of-season lines, overstocks, seconds and other lines which for trading reasons retailers can no longer allow to take up valuable space in full-price high street shops. Because goods are sold at discount prices, factory outlet centres attract visitors from a wide area who like bargain-hunting and who are out for a leisure-orientated day trip.

The development of factory outlet centres

Individual factory shops have been quite common in Britain for many years. The first 'embryonic' factory outlet centre was developed out of some conventional retail units around an original pottery factory shop at Hornsea in the East Riding of Yorkshire. The Hornsea Freeport Village opened in 1992, followed in 1993 by the first purpose-built centre, Clarks Village, at Street in Somerset. Both centres also have significant leisure attractions (Jones, 1995a). By the end of 1999 there were 29 factory outlet centres operating in Britain, mostly in England. A further 11 schemes were under construction or had planning permission. It has been predicted that factory outlet centres in Britain will soon reach saturation level because the market will only be able to accommodate perhaps 50 centres with large, overlapping catchment areas. It will also be increasingly difficult to obtain planning consents.

Factory outlet centres had their origins in the United States in the 1970s and they have become an increasingly important element in the American retail scene. Various lessons have been learnt by US operators over the last 20 years:

- It is not possible to develop successful outlet centres close to major retail centres because manufacturers are too sensitive about the effects on sales of full-price items in those centres.
- Visits to factory outlet centres are infrequent, necessitating good access to a large catchment population.
- There is a high degree of linkage between visits to factory outlet centres and other leisure-related activities (Booton, 1994).

All of these lessons are relevant to the development of factory outlet centres in Britain. They are all arguments that have been used by promoters of schemes in seeking planning approval. However, according to Baldock (1996), four 'fictions' about factory outlet centres need to be pointed out if local authorities and town centre retailers are not to be misled, as shown in Table 7.1.

Government policy towards factory outlet centres is set out in PPG6:

For factory outlet centres, the issue for planning policy purposes is not whether goods are sold at a discount, but whether such retail

Table 7.1 The case for and against factory outlet centres

Claim	Response
1 They sell goods which are not available in town centres and are therefore not in competition with town centres.	Expenditure is finite and any expenditure at factory outlet centres will reduce the amount available for spending in town centres.
2 They capture the 'leisure spending' rather than the retail spending which goes to town centres.	There is a limit to the amount of clothes, shoes, etc. that people need or are willing to buy.
3 Trade attracted to factory outlet centres benefits the centre of the town where the factory outlets are located.	Most developments are located on the edge of towns or out-of-town, where there is virtually no potential to generate linked trips.
4 When a planning application is made, the supporting retail impact study is objective and accurately indicates the likely impact on nearby town centres.	Developers' impact studies and other supporting information are commissioned to make a case in favour of the scheme but they are not objective documents.

developments would divert trade in comparison goods from existing town centres, whether they would be accessible by a choice of means of transport and, in particular, whether they would have a significant effect on overall car use. These centres would draw customers from a wide catchment area, predominantly by car, and as a result are unlikely to be consistent with the criteria in this guidance, unless these issues can be satisfactorily resolved.

(DoE, 1996, para. 3.9)

There has been a hardening of government attitudes since 1996. The Planning Minister, Richard Caborn, said in a speech in November 1998:

There is nothing special about factory outlet centres in planning terms. They are no more than a collection of shops selling comparison goods which seem to want to avoid existing centres, but want to serve a regional market. There is absolutely no reason why the stores in such centres could not operate in the high street – apart from the desire to avoid competition. We wish to encourage competition – we want to see these stores in town centres. This may put a question mark over the factory outlet centre format in so far as it seeks out-of-centre locations.

(DETR, 1998)

Evidence of the impact of factory outlet centres

There is so far a lack of evidence about the impact of factory outlet centres. On one hand, it may be argued that such outlets, because of their potentially large catchment areas, have a diluted impact on particular town centres, but on the other hand the type of goods normally offered (fashion, china, gifts, etc.) are also those sold by the most vulnerable town centre traders (Holt, 1998). The concern about the effects of a factory outlet boom is quite understandable. Since most factory outlet centre proposals are located out-of-centre there is a risk of diversion of comparison goods expenditure, though the potential effect will vary from town to town. At Bicester, for example, the outlet centre is next to a Tesco superstore and the combined retail floorspace is greater than that available in the town centre (Jones and Vignali, 1993).

Developers of factory outlet centres argue that the levels of trade impact will be low, generally less than 3 per cent of comparison goods trade (Jones, 1995a). Retail impact studies that have been carried out on factory outlet centre proposals also suggest that trade diversion from existing centres would be low. In the Hartlepool Retail Study carried out for Hartlepool Borough Council by England & Lyle in 1996, it was estimated on the basis of household and shoppers survey information that the Jacksons Landing factory outlet centre in the Marina had resulted in a trade loss of only 3 per cent from Hartlepool town centre. This was a surprisingly low impact considering the location of Jacksons Landing which was very close to the town centre, but it is explained by evidence that the trading performance of the factory outlets was relatively low and that two-thirds of the trade was drawn from outside Hartlepool. A similar situation is true of Hornsea Freeport Village where the impact is dispersed over a wide area and nearby centres have not been adversely affected.

A survey in 1998 by England & Lyle of most of the factory outlet centres operating at that time showed that:

- most have opened since 1994
- they mostly range in size from 4,000 to 20,000 square metres gross
- most are out-of-centre but some are edge-of-centre
- they are generally on brownfield rather than greenfield sites
- they are located mostly alongside leisure and A3 uses, and in some cases next to foodstores
- very few studies have been carried out post-opening
- the views of local authorities are generally positive; most local authorities reported no evidence of adverse impacts on town centres and benefits of attraction of visitors to the locality.

Colliers Erdman Lewis has carried out a research study into the economic effects of factory outlet centres on the fashion retail sector. The focus was on fashion because factory outlets are predominantly occupied by fashion retailers. The results showed that:

- trade diversion from fashion shops in town centres was low and was partly offset by spin-off retail expenditure from visitors drawn from outside the usual catchment of the town centre.
- a survey of fashion shops in town centres located close to established factory outlet centres showed few adverse effects.
- in all the town centres studied, fashion shop numbers have remained virtually static since the nearby outlet centre opened, while fashion floorspace actually increased (Doidge, 1999).

There is a contrast here between the findings of this study and the experience at Bicester where the number of town centre shops selling clothing and footwear has fallen by about a third since the factory outlet centre opened (Baldock, 1998). This may be because Bicester is one of the most successful factory outlet schemes. There are cases of high predicted impacts of proposed factory outlet centre developments. At Weston-super-Mare, Hillier Parker forecast that a proposed factory outlet centre would have an impact of 16 per cent on the town's durable goods sales and 32 per cent on clothing and footwear sales. In Stirling it was forecast that an outlet centre would have an impact of 15 per cent on town centre durable goods sales and 28 per cent on clothing and footwear (Baldock, 1998).

Quite apart from their quantitative effects, the Environment Committee was concerned with the danger that factory outlet centres are allowed to sell general lines rather than seconds or end-of-line goods, so making them indistinguishable from town centre shops. This has already happened in the USA. While factory outlet centres have been very successful there, Jones (1995a) points out that the retail structure, urban fabric and many cultural values are different in Britain. In the USA the market dominance of named brands has been crucial but in Britain the consumer has a much stronger loyalty to retailers, e.g. Marks and Spencer, than to particular product brands. Therefore, in Britain the appeal of factory outlets is likely to be less important. Also, while there has been a major decline and decay of town and city centres in the USA, there is still a viable network of town and city centres in Britain where it is possible to obtain competitively priced clothes and household goods.

Policy towards factory outlet centres, now set out in PPG6 and ministerial statements, seems to have been firmly established in the landmark decision on the Cotswold Outlet Village, a large factory outlet centre proposal at Tewkesbury in Gloucestershire in 1996. The inspector recommended in favour of the scheme but it was rejected by the Secretary of State. In terms of impact, even the worst-case assessment failed to demonstrate a significant quantitative impact on any neighbouring centre. In fact, the Secretary of State highlighted the inherent uncertainties and dangers of relying too heavily on potentially unsafe impact studies. He agreed with the inspector that clear evidence was lacking that the Cotswold Outlet Village would have a sufficiently harmful impact on the vitality and viability of any existing town centre to justify refusal of planning permission on these grounds. But on wider PPG6 issues the Secretary of State found that the application failed the key tests on accessibility and travel impact.

It is likely that PPG6 and PPG13 issues will form the basis for future opposition to factory outlet centres because the preferred locations are near motorway junctions and these centres aim to attract shoppers from a very large catchment area (Moss and Fellows, 1995). Such issues determined the outcome of the controversial proposal for a factory outlet centre next to the A1(M) junction at Bowburn near Durham City in 1995. Following the Tewkesbury decision, the application for the Bowburn scheme was withdrawn. It is likely that fewer factory outlet centres will be built than was anticipated a few years ago, that they will be closer to town centres, and they will occupy brownfield rather than greenfield sites (Fernie S., 1996).

Regional shopping centres

The development of regional shopping centres

PPG6 defines 'regional shopping centres' as out-of-town centres generally over 50,000 square metres gross retail area, typically enclosing a wide range of comparison goods. Such centres have been common in North America for many years (see Chapter 9) but are relatively new in Britain. Early proposals for the development of regional out-of-town shopping centres in Britain dating from the early 1960s met with vigorous opposition from central and local government. An application for a regional centre of 100,000 square metres at Haydock, between Manchester and Liverpool, was refused in 1964 by the Ministry of Housing and Local Government, following a planning appeal. The main reason for refusal was the projected impact upon existing town centres in the region (Guy, 1994b).

The first regional shopping centre at Brent Cross was not approved until 1968 and did not begin trading until 1976. Even then, it was not strictly an out-of-town development and fitted into a gap in the traditional retail hierarchy in north-west London. The early 1980s saw a renewal of interest by developers in regional shopping centres, encouraged by a relaxation in government attitudes towards out-of-town retailing and the creation of enterprise zones (Guy, 1994a). New attitudes were bound up with the enterprise culture and a *laissez-faire* approach to retail planning by central government (Davies and Howard, 1988).

Four out-of-town regional centres were built in the 1980s; the Metro Centre was the first, followed by Merry Hill, Meadowhall and Lakeside. A further five were built in the late 1990s. At the time of writing the centres in existence are listed in Table 7.2.

Table 7.2 Regional shopping centres

Centre	Floorspace (sq. m. gross)	Opening date
Metro Centre, Gateshead	145,000	1986
Merry Hill, Dudley	130,000	1989
Meadowhall, Sheffield	120,000	1990
Lakeside, Thurrock	120,000	1990
White Rose Centre, Leeds	60,000	1997
The Mall, Cribbs Causeway, Bristol	70,000	1998
Trafford Centre, Manchester	120,000	1998
Bluewater, Dartford	155,000	1999
Braehead, Glasgow	90,000	1999

Evidence of the impact of regional shopping centres

PPG6 recognises that regional shopping centres can have a substantial impact over a wide area and can severely harm town and city centres. It states that in most regions there is unlikely to be scope for an additional regional centre without adversely affecting the vitality and viability of existing centres. There was a great concern among local authorities in the 1980s that regional centres would lead to a substantial loss of trade from town and city centres, leading to a major reduction in retail representation and investment confidence in such centres (Jones, 1989). The Environment Committee took the view that the benefits of regional centres have been counterbalanced by the fact they have led to 'cannibalism' of retail trade and employment from existing centres. But the committee felt that since the effects of regional centres vary considerably it is difficult to generalise about their impact. The key issues appear to be the strength of existing centres and the extent of the regional centre's catchment area.

Regional centres have attracted considerable research interest. Extensive research was carried out on the impact of the Metro Centre by OXIRM over a five-year period (1986 to 1991), focusing on its impact on established centres before and after surveys were undertaken. They showed that about 12 per cent of trips which would have been made to Newcastle city centre were being made to the Metro Centre (Howard, 1989). Although in its first five years the Metro Centre took a substantial share of retail sales in north-east England, its impact was not focused directly on Newcastle city centre but spread widely across a broad catchment area.

> Impact was experienced through a hastening or reinforcement of trends already working in retailing and shopping centres in the region. These trends included consumer preference for larger or more specialised or more convenient shopping centres and to use their mobility to visit them. Impact has been most adverse, not in the largest centre or even the centres nearest to the Metro Centre, but in the weaker centres and the weaker parts of centres.
> (Howard and Davies, 1993: 148)

Research on the impact of Merry Hill shows that the scheme has been a success commercially but it has had a serious effect on existing centres (Roger Tym and Partners, 1993). The impact has been most concentrated on nearby centres, particularly Dudley, with:

- a loss of market share
- a decline in floorspace
- a high vacancy rate
- a collapse in investment confidence in the town centre
- a loss of most of the multiple retailers from the town.

Over half of the 70 or so multiple retailers present in Dudley in 1986 subsequently left the centre, including Marks and Spencer, Littlewoods, British Home Stores, C & A, Burtons, Currys, Halfords and Mothercare, as well as Sainsbury's. Some firms moved directly into Merry Hill, while others left after the initial impact began to affect shopper visits to the town centre. Most of the vacated shops were reoccupied but mainly by low quality discount and variety stores. The centre has clearly suffered a loss of vitality and viability (Guy, 1994a). It is thought that Merry Hill has had a severe impact on Dudley because it does not enjoy good accessibility and it has a relatively limited catchment area for a development of its size. Stourbridge has also suffered a very significant impact. Although Merry Hill has diverted some trade away from the two largest comparable centres in the West Midlands – Birmingham and Wolverhampton – the conclusion was that these centres remained buoyant (Roger Tym and Partners, 1993).

OXIRM was also involved in a five-year programme to monitor the changes resulting from Meadowhall. The findings show a very extensive catchment area, with more than 30 per cent of trade being drawn from over 30 minutes drive away, compared with 25 per cent for the Metro Centre. The average expenditure made on trips to Meadowhall is much higher than for existing centres. Initially there was evidence of a serious impact on Sheffield city centre in terms of a decline in trade and a high vacancy rate. Early estimates of trade diversion made in 1986 argued that there would be only a 4 per cent trade loss in Sheffield city centre, but an updated report in 1988 increased the estimate to 12 to 13 per cent. This is similar to an impact assessment by Sheffield City Council which calculated a trade diversion of 14 to 15 per cent (Williams, 1991). By 1996 there were signs that Sheffield was re-establishing itself as a major shopping centre. In its unitary development plans the council wanted a complete embargo on any more out-of-town retailing at Meadowhall, but the inspector thought this was too restrictive and modified the policy to prevent any more out-of-town retailing at Meadowhall unless it can be shown that it would not have an adverse impact.

The Lakeside centre has not been subject to as much investigation as the other first generation regional centres. Lakeside was approved after an inquiry despite the fact that it was accepted there would be a substantial loss of trade in Grays town centre, 10 minutes drive away. There is evidence that the vitality and viability of Grays has been adversely affected. The four regional centres built in the 1980s are the only ones out of a total of about 50 which were originally proposed. Some were approved but subsequently found not to be viable schemes, but most were refused on retail impact or policy grounds. For instance, the Centre 21 scheme near Leicester was dismissed on appeal in 1982 but only after the Secretary of State had overturned the inspector's recommendation to allow it. The inspector accepted figures of 17 per cent impact on comparison goods expenditure in Leicester city centre, 13 per cent in Hinckley and 12 per cent in Loughborough. The Secretary of State decided that these impacts would seriously affect the vitality and viability of the centres in question.

Probably the most controversial regional centre proposal has been the Trafford Centre at Dumplington in Manchester. The original planning application was made in 1986 and, after two inquiries, planning permission was issued in 1993. But following appeals to the High Court and the Court of Appeal, the consent was quashed on legal grounds in 1994. The House of Lords overruled this decision and granted planning permission in 1995.

Also controversial was the proposal for a large-scale extension to the Merry Hill Centre which was called in by the Secretary of State in 1995. The extension would have increased the floorspace to almost 200,000 square metres. The proposal was subject to a public inquiry in 1996 at which impact arguments were paramount. At the inquiry significant impacts were predicted on other centres, especially Dudley, even by the owners of the centre. The inspector concluded that impact was likely to be in excess of 20 per cent on Dudley and approaching this level on Stourbridge and Brierley Hill. The proposed extension was refused by the Secretary of State in 1997.

Proposals were made in 1999 for a 35,000 square metre extension of the Metro Centre which includes a large Debenhams department store, replacing the Asda superstore which has relocated to a freestanding site on the adjacent retail park. After an inquiry the extension was approved by the Secretary of State.

Concerns about the impact of regional shopping centres on established centres are reflected in appeals by retailers to the valuation office for reductions in rateable values resulting from losses of trade to regional centres (Davey, 1999a). Historically reductions in rateable values have been agreed as shown in Table 7.3.

Retailers in Wakefield, Dewsbury and Morley have also been awarded reductions of up to 10 per cent in their rateable values because of the impact of the White Rose Centre. More recently appeals have been lodged in towns near to the Trafford Centre, Bluewater and The Mall, Cribbs Causeway (Davey, 1999a).

Research published in *Property Week* shows that retailers in Manchester city centre experienced a drop in sales of 11 per cent in the year to May 1999, while in Stockport sales dropped by 13 per cent. Other declines were reported in Bolton and Altrincham. These declines are attributed to the opening of the Trafford Centre.

It is generally accepted that, now the regional shopping centres which were in the pipeline have been built, government policy has effectively brought an end to the development of such centres. PPG6 says any new proposals should be brought forward through the development plan process, but it is difficult to see how they could satisfy the government's sustainable development strategy and its objective to sustain and enhance the vitality

Table 7.3 Reductions in rateable value attributable to regional shopping centres

Centre	Area	Reduction in rateable value
Metro Centre	Newcastle	up to 12%
	Sunderland	7.5%
	Gateshead	up to 30%
	Durham	6%
Merry Hill	Brierley	20%
	Stourbridge	10%
Meadowhall	Sheffield	15%
	Rotherham	10%
Lakeside	Grays	17.5%
	Romford	11%
	Barking	15%
	Basildon	12.5%

Source: *Property Week.*

and viability of town centres. Braehead is likely to be the last of the regional shopping centres.

Summary

The evidence of retail impact varies between different types of retail development — foodstores, retail parks, factory outlet centres, and regional shopping centres. There is intense competition between superstores, particularly the four major food/grocery retailers and the grocery market is becoming saturated with about 1,100 superstores in Britain. The rapid growth of superstores in the 1980s and early 1990s led to concerns about their impact on existing centres. The initial fears about impact have gradually declined but there are still issues about the effects of superstores on small towns. Research on large foodstores in market towns and district centres shows that they can and have had an adverse effect on these centres. In the early years of superstore development, trade diversions of up to 10 per cent of convenience turnover were generally thought to be acceptable, but later the threshold of acceptance tended to rise to 15 per cent. A very high predicted level of impact will invariably lead to a refusal of an appeal or call-in. However, a convenience trading impact of less than 10 per cent may equally be unacceptable if a centre has a low level of vitality and viability. There is less concern about the impact of discount foodstores, but they can sometimes have significant impacts on vulnerable centres. Cumulative impact issues have also to be considered, in some cases, for superstores and discount foodstores.

There has been an evolution from the development of retail warehouse 'sheds' selling non-food bulky goods to large, modern retail parks in response to the rapid growth of consumer demand in this market. There are now about 500 retail parks in Britain. Most retail parks are out-of-centre but government policy favours edge-of-centre locations which have benefits in terms of linked trips, accessibility and regeneration of brownfield sites. The impact of retail parks depends on their size and the range of comparison shopping that they offer. Local authorities have tried to control the range of goods sold by means of planning conditions or legal agreements. There is little evidence of adverse trading impacts on established shopping centres. Impacts are often qualitative rather than quantitative. Many appeals have been allowed because impacts were considered to be insignificant, and many retail parks have been built without any adverse impact. However, some very large proposals have been refused permission on impact grounds.

Factory outlet centres are relatively new in Britain, with only 29 built and another 11 under construction or approved. They have a large drawing power across a regional catchment, providing a leisure as well as a retail function. Government policy is opposed to any further out-of-centre factory outlet centre developments, mainly because they attract mostly car-borne trade, rather than for impact reasons. So far there is a lack of evidence about the impact of factory outlet centres, but generally the existing developments have not had significant impacts on town centres. The scope for further developments of this type is limited and the trend is likely to be towards more central locations and brownfield sites.

Four out-of-town regional shopping centres were built in England in the 1980s and four more in the 1990s, with another one in Scotland. PPG6 notes that regional shopping centres can have a substantial impact over a wide area and can severely harm town and city centres. The impacts vary considerably depending on the strength of existing centres and the extent of the regional centre's catchment area. Merry Hill and Meadowhall have had a generally greater economic impact than the Metro Centre. Many proposals for regional shopping centres have been refused on impact grounds, as has the proposed extension to Merry Hill. Rateable values have been reduced in several towns near to regional centres as a result of losses of trade by retailers in these towns. Government policy has effectively brought an end to the development of regional shopping centres.

8

RETAIL IMPACT IN THE PLANNING PROCESS IN BRITAIN

Earlier chapters have referred to the deficiencies in the methodology of RIA and the need for an improvement in the application of RIA in practice. The best practice recommendations in Chapters 5 and 6 are aimed specifically at bringing about these improvements. Some of the evidence underlying the problems associated with current approaches is presented in this chapter. First, it indicates the views of local authorities on RIA and the ways in which the process needs to be improved. Second, a review is made of the influence of retail impact factors in decisions on planning appeals and call-ins on proposals for retail developments. The significance of government policy on planning decisions affecting retail development is clear. The findings of this chapter are important in showing why the recommended approach to best practice is essential to ensure that the application of RIA is consistent with the current government policy on retailing.

Views of local authorities

As part of the research for the author's PhD thesis, a postal questionnaire survey of local authorities was carried out in 1995. The survey was devised to explore the link between RIA and the policy climate in which it is carried out in Britain. In particular, it aimed to examine how local authorities respond to the pressure for new retail development, and their attitudes towards retail impact issues in the planning process. The content of the survey covered three main areas which are particularly relevant to the analysis of these issues:

- *Policy* – on the vitality and viability of shopping centres and on major developments outside existing centres
- *RIAs* – the requirement to undertake an RIA, and opinions and comments on RIAs that have been carried out
- *Experience of major retail developments (for district councils only)* – including recent planning applications, and evidence of harm to the vitality and viability of town centres and the cumulative impact.

162

Questionnaires were sent to a total of 165 local authorities in England, Scotland and Wales. All counties existing at that time were included in the survey because of the importance of strategic policy on shopping and the technical guidance that counties provide to districts on retail planning matters. A sample of one in four districts was selected to keep the survey to a manageable size. The approach used was designed to provide a sample which is representative geographically and by size of district in terms of population. The survey was carried out and completed before the reorganisation of local government in Scotland, Wales and some parts of England in April 1996. In Scotland and Wales the regional and county councils were abolished at that time and new unitary authorities created. A very good response rate of 89 per cent was achieved, and the results are representative of local authorities across the country, for both counties and districts.

The local authorities were asked whether their structure plan, local plan or unitary development plan includes policies to support the vitality and viability of existing town/city centres. The survey shows that within the development plan framework there is firm support for town and city centres and policy for dealing with major or large-scale retail development outside town centres, both in structure plans and local plans. Not all development plans fully reflected the latest government policy on retailing because they were prepared at different times and it is difficult for local authorities to keep pace with the rapid changes that have been made in PPG6 guidance. However, development plans have taken account of changes in government policy guidance as they have become modified or reviewed, and there are signs that policy at the local level is taking account of the latest guidance in the revised PPG6 concerning the promotion of town centres and the sequential approach to selecting suitable sites for retail development. There is no question of there being standard policies for retail development; local circumstances will influence the nature of policy at the local level, but policy must be consistent with the national guidance.

The survey included a series of questions on local authorities' views about RIA. The questions covered the following issues:

- whether there is policy backing for local authorities to ask for an RIA
- whether local authorities seek independent audits of RIAs
- the general opinions on the quality of RIAs
- the ways in which RIAs could be improved.

The local authorities were asked whether they have a requirement for applicants proposing major retail development to submit an RIA. Most local authorities (60 per cent) do have a requirement for applicants to prepare an RIA as part of their planning application submission. The extent to which this is a requirement varies between counties and districts. A relatively low proportion of counties (37 per cent) expect RIAs to be produced, but it has to be recognised that counties are not the local planning authority and so cannot insist that applicants produce a retail impact statement. Several counties pointed out that it is implicit in the wording of structure plan policy that in practice an RIA would normally be expected. However, the majority of district councils require applicants to produce an RIA, and this is usually laid down in the UDP or local plan policy.

The local authorities were then asked whether they commission independent audits of applicants' RIAs. It is becoming increasingly common for local authorities to obtain a second opinion on the impact of a proposed development by commissioning consultants themselves to make an independent audit of the RIA that has been submitted. From the survey, the district councils' responses were evenly split between those who do and those who do not opt for an independent audit.

The local authorities were also asked about their opinion of RIAs submitted by applicants, indicating on a five-point scale whether they are very good, good, average, poor or very poor. Almost half of the local authorities who responded to this question thought that the standard of retail impact statements was average. This is perhaps to be expected but it is significant that more local authorities thought they were good rather than poor. Districts have a generally higher opinion of RIAs than counties who appear to have a more cautious attitude, which is reflected in the fact that a quarter of the counties said that RIAs vary in quality. It was also mentioned that the quality depends on the applicant or consultant, and that the larger the development, the more comprehensive the retail impact study. The results show that, according to the views of local authorities, the overall quality of RIAs needs to be improved.

The local authorities were asked in what ways the assessment of retail impact of proposed developments could be improved. The answers are given below in rank order. There is a need for:

- better/more up-to-date retail statistics at the local level
- RIAs to be more independent/objective/impartial
- surveys of local shopping patterns/catchment areas
- less bias in favour of the development
- more co-operation between applicants and local authorities
- more emphasis on vitality and viability indicators
- more emphasis on transport/traffic/environmental issues
- a standard/consistent methodology
- more information from retailers, e.g. on turnover
- a range of scenarios/assumptions (sensitivity)
- consideration of the cumulative impact
- post-development studies to check assumptions/accuracy.

Almost all local authorities expressed an opinion on this topic. Two major conclusions are evident from the responses:

*   The availability and quality of information used in RIAs is clearly inadequate. Over a third of all comments referred to the need for better information at local level or the need for local surveys of shopping patterns.
*   Local authorities are also very critical of the bias and lack of impartiality shown in RIAs produced by consultants. Almost a third of all comments mentioned this as a problem.

Counties expressed a particular concern about the problems of retail information, with 40 per cent of their responses on this point. Districts are also concerned about data problems but generally they are more concerned about the objectivity of RIAs. It is districts that have to understand and interpret the findings of retail impact statements to help in making judgements about planning applications. Most of the comments from districts are about the practical issues of how accurate retail impact statements are.

A series of questions were included in the survey asking district councils about their experience of decisions on major retail developments and their views on the impact that retail developments have had on town centres. Councils were asked whether any retail developments which have taken place have harmed the vitality and viability of a town centre. Forty-two per cent of respondents answered 'yes'. This is a significant proportion and it suggests that the impact of new shopping developments on existing centres is of great concern to local planning authorities. However, the results reflect the opinions of planners responding to the survey rather than firm evidence that impact has occurred. Most local authorities that indicated that there has been harmful impact referred to the effects of out-of-town superstores and retail parks on town centres. Two specific factors were identified by several local authorities: the impact of particular large out-of-town centres such as Merry Hill and Lakeside on nearby town centres, and the effect of retail developments which have taken place outside the local authority area, i.e. competition between neighbouring towns.

As part of their research for DETR into the impact of large foodstores on market towns and district centres, CB Hillier Parker also carried out a postal survey of all local authorities in England and Wales in 1996. A response rate of 43 per cent was obtained, producing slightly more responses in total than the author's sample survey in 1995. Over half of all respondents said they were facing pressure for further foodstore development. Almost 75 per cent of local authorities usually require an RIA when considering applications for new foodstore development. This compares with a figure of 60 per cent in the author's survey the previous year. The increase reflects the shift in government policy which occurred in the draft revision of PPG6 in 1995. It appears that there is no generally accepted approach by local authorities to the requirement for RIAs; there is a lack of consistency between authorities when considering the effects of a new retail development. However, local authorities are predominantly concerned with the impact of out-of-centre locations, and are more likely to request an RIA in the case of out-of-centre proposals (CB Hillier Parker, 1998).

Sixty-nine per cent of local authorities facing pressure for further foodstore development use external consultants to undertake RIAs. It is seen to be a relatively specialist field in which local authorities generally do not have expertise. However, almost 45 per cent of local authorities consider the current RIA methodologies are inadequate. A general concern was expressed that the key variables could easily be manipulated to endorse a particular viewpoint. Doubts were expressed about the objectivity of many applicants' RIAs (CB Hillier Parker, 1998). These findings confirm the conclusions of the author's survey and highlight the need for improvement in the methodology and application of RIA, and the need for advice on best practice.

Retail impact factors in planning appeal decisions

Although most proposals for major retail development are decided by local authorities without going through the appeal process, a substantial number of proposals are decided on appeal by planning inspectors. Inspectors have to deal with conflicting evidence produced by the parties to make a judgement about the merits of the proposal. An analysis has been made of records from the computerised planning appeals database (COMPASS) of all appeal decisions on retail developments (including call-ins) in England, Scotland and Wales from 1988 onwards, giving details of the type of development, location, factors involved in the decision, and date of decision. It is important to see how the pattern of decisions has been influenced by changes in the national policy context. Therefore attention is focused on the period since the introduction of PPG6 in 1988 and its subsequent revisions in 1993 and 1996. The analysis included appeals allowed or dismissed, the number of appeals decided each year, and the reasons for the decision.

Since 1988 a total of about 800 planning appeals for retail developments (including call-ins) have been determined in Great Britain. The only exclusions that have been made from the COMPASS database are minor extensions, variations of conditions, access, etc. A distinction is made between food and non-food retailing. Summary statistics of appeals allowed and dismissed are shown in the graph in Figure 8.1.

The number of appeals declined from a peak in 1988, at the height of the property boom, when the number rose to more than 120 per annum. This was followed by a sharp drop to only about 40 per annum in the period 1991–92. As the economy picked up again, a further increase in appeals

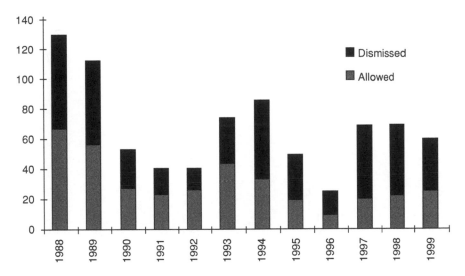

Figure 8.1 Number of appeals for retail development, Great Britain

decided occurred in 1993–94 before another decline to 1996. The numbers have increased again in 1997–99. The trends in appeals allowed and dismissed follow the same general pattern but there is more fluctuation in the numbers dismissed. To understand the reasons behind these trends it is necessary to look at the pattern of proposed food and non-food retail developments, which is shown in Table 8.1.

For the purposes of analysis it is useful to group the data into three time periods: 1988–93 (PPG6 was revised in July 1993), 1994–96 (PPG6 was revised again in June 1996), and 1997–99. In the period up to 1993 there were a similar number of appeals for food and non-food retail developments. However in 1994–96 most appeals were for foodstores. Pressure for foodstores remained at about the same level as previously but proposals for non-food developments fell to a much lower level, only half of that for food-stores. Since 1996 the number of appeals on foodstore proposals has increased steadily but the number of appeals for non-food developments has declined after a peak in 1997. These trends can be interpreted as signs that the food retailing market has remained buoyant but the market for non-food retailing reflects an uncertain economic climate.

The proportion of appeals allowed in these three time periods is as shown in Table 8.2.

Overall, the proportion of appeals allowed has declined steadily since 1988. For food and non-food stores the pattern is similar, but there has been a higher success rate for non-food developments than foodstores but this gap has narrowed in recent years. PPG6 was introduced at a time when an

Table 8.1 Appeals allowed and dismissed, Great Britain

Appeals allowed

Year	Food	Non-food	Total
1988	26	40	66
1989	22	36	58
1990	13	16	29
1991	13	11	24
1992	19	7	26
1993	31	14	45
1994	14	21	35
1995	18	5	23
1996	10	0	10
1997	9	12	21
1998	9	15	24
1999	17	9	26
Total	201	186	387

Appeals dismissed

1988	20	43	63
1989	22	31	53
1990	16	8	24
1991	11	5	16
1992	12	2	14
1993	22	6	28
1994	36	14	50
1995	16	10	26
1996	12	2	14
1997	16	31	47
1998	24	20	44
1999	23	10	33
Total	230	182	412

Table 8.2 Proportion of appeals for retail development allowed, Great Britain (per cent)

	Food	Non-food	Total
1988–93	55	57	56
1994–96	40	50	43
1997–99	36	37	36

unprecedented number of appeals for major retail development were being determined and there was a marked reduction in appeals following the new policy guidance. The revised PPG6 in 1993, with its more restrictive policy stance, led to a tougher line being taken on appeals and an increase in the rate of dismissals. This tougher line has continued since the 1996 revision of the guidance. In 1997 less than a third of all appeals on retail developments were allowed but by 1999 the success rate had risen again to over 40 per cent. It is likely that many schemes which had no hope of success in the light of the new government policy were not pursued to appeal.

An analysis has also been made of the factors that have been referred to by inspectors in determining appeals/call-ins for major retail development. These factors are summarised in the COMPASS database and the records have been examined to establish the reasons why appeals have been allowed or dismissed. The details are shown in Table 8.3.

Table 8.3 Factors involved in decisions on major retail developments, Great Britain

Factor	1988–93	1994–96	1997–99
Appeals allowed			
Need accepted	18	8	10
No central sites available	3	2	0
No significant retail impact	112	39	34
No conflict with policy	34	5	7
Amenity/environment issues	49	12	9
Traffic/access/parking issues	56	5	9
Industrial land issues	52	8	8
Regeneration benefits	11	3	5
Employment benefits	22	9	4
Linked trips	0	2	13
Reduction in car travel	0	12	5
Accessible by public transport	0	11	22
Meets the sequential test	0	10	26
Total	*357*	*126*	*152*
Appeals dismissed			
Effect on town centre investment	10	4	4
Impact on other centres	61	25	54
Cumulative impact	13	5	11
Contrary to policy	38	20	21
No evidence of need	20	7	21
Amenity/environment problems	58	23	25
Traffic/access problems	31	11	12
Car parking/servicing	7	0	2
Over-development of site	10	0	0
Loss of industrial land	48	16	28
Loss of housing land	2	3	3
Loss of agricultural land	6	2	0
Not accessible by public transport	1	15	16
Fails the sequential test	0	12	53
Increase in car journeys	0	9	12
Total	*305*	*152*	*262*

Appeals included are for new retail development and extensions. Excluded are appeals for variations of conditions, costs, advertisements, access and car parking.

The most important factors leading to proposals being approved since 1988 are no significant retail impact, industrial land issues, amenity/environment issues, and traffic/access/parking issues.

The lack of any significant retail impact is by far the most important factor in appeals allowed, being mentioned in about a third of cases. In the latest period, 1997–99, lack of retail impact still accounted for over 20 per cent of the reasons for appeals being allowed. What is also significant about the most recent experience is the emergence of the new factors introduced by PPG6. Satisfying the sequential test and accessibility by public transport are now the next most important factors influencing approvals of proposed retail developments.

In the case of appeals dismissed, the most important determining factors in appeals/call-ins since 1988 have been the impact on other centres, amenity/environment problems, loss of industrial land, being contrary to policy, traffic/access problems, and failing the sequential test.

Adverse retail impact was referred to in about 20 per cent of cases dismissed between 1988 and 1999. In 1997–99 the proportion was still over 20 per cent. Therefore, retail impact factors are slightly less important as reasons for dismissing appeals than for appeals allowed. The most significant change in determining factors in refusals is appeals which fail the sequential test, which is now the next most important factor and is almost as important as retail impact.

Since 1997 the main factors in foodstores allowed/approved on appeal have been meeting the sequential test, the lack of evidence of impact and good accessibility by public transport. Decisions on appeals dismissed/refused have been mainly on the grounds of failing the sequential test and evidence of serious retail impact. The key tests in PPG6, therefore, have become the major determining factor in appeals on foodstore developments. In the case of retail warehouses and retail parks the most important factor in proposals allowed or approved in recent years has been the lack of evidence of retail impact. Inspectors were satisfied that the proposals would not harm the vitality and viability of nearby centres. However, meeting the sequential test has become a much more important factor in the last few years, in line with the revised PPG6. There are also signs that the potential for linked trips and a reduction in car journeys are also becoming more significant factors in decisions on appeals/call-ins for retail warehouses and retail parks. Other key factors in approvals are good public transport and accessibility, and the fact that the loss of an industrial site is not significant. An increasing proportion of appeals allowed for retail parks are edge-of-centre schemes, but refusals are predominantly for out-of-centre schemes. The trend is firmly that edge-of-centre developments have a much higher chance of success on appeal/call-in. The most important factors in those proposals for retail parks which were dismissed/refused have been the loss of industrial/employment land and concern about harm to the vitality and viability of centres. But in the last few years the influence of the revised PPG6 has increased, with more proposals being rejected because they fail the sequential test or because sites are not accessible by public transport.

Summary

A postal questionnaire of local authorities in Great Britain shows that there is firm support for town and city centres and policy for dealing with out-of-centre retail developments. Most local authorities have a requirement for applicants to submit an RIA as part of their proposals for retail development. It is becoming increasingly common for local authorities to commission consultants to prepare independent audits of RIAs submitted by applicants. The local authorities believe that the overall quality of RIAs needs to be improved. They are particularly critical of the quality of information used and the bias and lack of impartiality shown in RIAs. These findings are confirmed by other, more recent, survey information.

A substantial number of proposals for retail development are decided on appeal by inspectors (and reporters in Scotland). The number of retailing appeals in Great Britain has fluctuated in the period 1988–98 and it is currently at a much lower level than in the late 1980s. The proportion of appeals allowed for retail development has declined steadily since 1988. The revised PPG6 in 1993 led to a more restrictive policy stance, and this tougher line has been strengthened since the 1996 revision of the guidance. Now only about 40 per cent of all appeals which are retail developments are allowed. The most important factor in proposals being approved has consistently been the lack of significant retail impact, but accessibility by public transport and the sequential test are becoming increasingly important. Appeals are dismissed mostly because proposals have an adverse retail impact and/or fail the sequential test. The pattern is similar for foodstores and non-food retail parks. The key tests in PPG6 have become the major determining factors in appeals on both foodstore and retail park developments. Edge-of-centre proposals have a much higher chance of success than out-of-centre developments.

9

EXPERIENCE IN EUROPE AND NORTH AMERICA

Experience of dealing with pressures for large new retail developments and the response of the planning system to these pressures in Britain is now very well developed. This chapter looks at the experience of Europe, particularly western Europe, and North America in terms of the type of retail development that has taken place, the planning response to these developments, the evidence of retail impact, and the approaches to impact assessment. Europe and North America have been selected because they represent areas which have seen retail development take place on a large scale over recent decades, and which may be expected to provide lessons about approaches to retail planning in Britain. It is not an exhaustive coverage of retail development in these areas but simply a commentary on international experience as a basis for comparison with the more detailed examination of retail impact issues in Britain.

International comparisons are relevant to an understanding of new retail development in Britain and the way in which the planning system seeks to influence the retail development process. The following issues are particularly important in comparing Britain with other countries:

- What is the latest evidence of impact in different countries?
- How does impact vary between different types of development?
- How have governments responded to pressures for development?
- What methods are used to assess impact, and is there a commonly accepted 'preferred approach'?
- Are decisions based mostly on economic impact or are other factors, e.g. social and environmental, also considered?
- What lessons can be learnt for the application of RIAs in Britain?

Europe

Retail development in Europe

The growth of out-of-centre retailing in Europe is most evident in France. Over the last 30 years France has decentralised a large part of its retailing from town centres to edge-of-centre and out-of-centre sites (*Property Week*, 1999a). The hypermarket concept originated in France, initially by Carrefour, and there was a rapid growth in the late 1960s and 1970s. The number of hypermarkets rose to about 450 in 1981, then doubled again to over 900 in 1991. However, the rate of development in the 1980s was slower than in Britain where the number of superstores increased from 280 in 1981 to more than 730 in 1991.

The significance of hypermarkets in France can be attributed to the higher density of urban development compared with towns and cities in Britain. There was little room for expansion in the central areas and a lack of sites for new developments. Hence the trend towards out-of-town development (Mills, 1974). Virtually all large retail developments in France have taken place on greenfield sites. Until the advent of the hypermarket in France, independent retailers still dominated the grocery market (Davies, 1976).

Out-of-town regional shopping centres first appeared in Europe in France in the late 1960s and early 1970s. The Parly 2 centre (87,000 square metres) which opened in 1969 was the first of several fully enclosed centres to be built in the Paris region. These centres were anchored by branches of department stores already operating in the centre of Paris. Several such centres were built in peripheral locations around the major conurbations in the period up to the mid-1970s, some of them related to the development of new towns (Dawson, 1983). Most of the centres developed around Paris have now lost their department store anchors, to be replaced by hypermarkets (Reynolds, in Bromley and Thomas, 1993). These centres now dominate the French retail market, with 25 per cent of all retail activity (*Property Week*, 1999a). There are now ten factory outlet centres in France. It is second to Britain in the amount of factory outlet space already trading (*Property Week*, 1999b).

Germany's experience of retail change has been more similar to that of Britain although there are certain basic differences. The most important of these is the number of large discount and variety stores in out-of-town locations (Davies, 1976). The discount food sector is of great importance in Germany. Retailers such as Aldi and Lidl are among the highest turnover food retailers there. Several large out-of-town centres have opened. The first purpose-built regional shopping centre in Germany is CentrO at Oberhausen, which opened in 1996 (*Property Week*, 1999a).

Other parts of Europe have also seen a trend towards superstores and out-of-town centres. In Sweden there has been sustained pressure from the retail industry since the late 1960s for developments to be sited away from established city centres. Because of the geographical distribution of population, almost all out-of-town developments are located in the south of Sweden. As in Britain, there has been a decrease in the number of retail stores and an increase in the dominance of multiples, particularly in food retailing (Westlake and Forsburg, 1996).

In The Netherlands shopping centres exceeding 40,000 square metres sales area did not appear until the 1960s. The threat of these new shopping centres did not provoke changes in the planning machinery to safeguard town centres. It was the advance of out-of-town superstores and discount stores which was seen to be more serious. The first Dutch superstore began operating in 1968 but by 1971 there were some 50 superstores occupying an average of 2,300 square metres (Borchert, 1988).

The planning response

Planning regulations vary considerably across Europe. Most countries have planning laws that attempt to protect traditional town centres from the perceived threat of out-of-centre retailing. There tends to be a stricter control over new retail development than in Britain, with legal backing involving locally organised impact studies and regulatory committees (Guy, 1994b). There is now evidence that the tighter planning policies that apply in Europe are having an effect in slowing down retail development. The development of regional centres in France, Belgium, The Netherlands and Germany has been almost halted since the late 1970s, mainly by government legislation controlling virtually all retail development outside existing town and city centres (Guy, 1994a).

In France the planning process is governed nationally but controlled locally by the *départements*, to whom an application for a CDEC (planning permit) is sought. The rapid growth of hypermarkets and concern about their impact on established centres led to government intervention and control of the development process in the early 1970s. The Loi Royer in 1973 initiated a restrictive policy regarding out-of-town developments. It established a series of *départemental* planning commissions composed of representatives of locally elected politicians, retailers and consumers. These commissions were given the power to authorise or reject planning applications for large retail units and extensions to existing stores above certain size thresholds depending on the size of the commune in which development was to take place. The law allowed for an appeal procedure in which the Minister of Commerce and Crafts, advised by a national commission of a similar composition, was the final arbiter (Burt, 1985). The original threshold was 3,000 square metres gross floorspace or 1,500 square metres sales area in towns of more than 40,000 population, or more than 1,000 square metres in smaller towns (Delobez, in Dawson and Lord, 1985).

The effect of the Loi Royer was an immediate reduction in the rate of development of large shopping centres. But the pressure for the development of hypermarkets continued (Dawson and Lord, 1985). Whilst the legislation had some short-term and localised impact on hypermarket development, in the long-term the restrictive effects of the law were less apparent. By exploiting various loopholes and failings in the legislation, hypermarket openings have continued and the large retail groups have maintained their growth largely at the expense of the smaller retailer, whom the law was intended to protect (Burt, 1985).

A moratorium on the further development of out-of-town shopping was introduced in France in 1993 as a reaction to pressure from small shop-keepers and others concerned about the impact on traditional retail patterns (House of Commons, 1994). Amendments to the Loi Royer have now made it much more difficult for developers to obtain planning permission for large retail developments. The floorspace threshold has been revised downwards to 300 square metres. In addition, since the last amendment to the Loi Royer in 1996, any retail development in excess of 6,000 square metres requires not only CDEC approval but also must go through a public inquiry. The process is the same for in-town as well as out-of-town developments, and the effect has been to slow down the rate of retail development (*Property Week*, 1998).

There are clear differences between Britain and Germany in retail policies. As in France a large number of out-of-centre developments were permitted in the 1970s but increasingly controls were placed on development. Germany maintained a firmer control on large new retail developments in the 1980s than happened in Britain. The federal government has no direct role in assessing retail proposals but sets the ground rules through the national land-use regulations. Development plans provide a policy framework for the development of large retail stores, and detailed zoning plans are issued on the basis of the development plans. Germany has fiercely resisted the trend towards out-of-town development, seeking instead to find ways to make the old centres accessible and more attractive (Hall, 1988). In Germany the tendency has been to preserve town and city centres as the focus for shopping, making it very difficult to gain planning consent for any shopping centre development in or out of town.

There is no embargo on out-of-town retail developments but the acceptability of large out-of-town schemes depends on compliance with area planning regulations (*Property Week*, 1998). The legislative framework is through the BauG or Federal Construction Law which controls the location of new retail development. It is supplemented by the BauNVO or construction use regulation. The BauNVO stipulates that shopping centres and large retail outlets can only be established in certain areas, such as outer industrial or mixed-use zones, without special authorisation, or they must be integrated with existing shopping (TEST, 1989). It also lays down a threshold floorspace figure for new developments. The BauNVO has had several amendments to cope with continuing pressures for large-scale retailing developments. An amendment in 1986 lowered the threshold to 1,200 square metres. These restrictions on large stores and their location have encouraged the emergence of smaller specialised discount stores (Kulke, 1996) and have led to the location of superstores in inner city areas (Zentes and Schwarz-Zanetti, 1988).

The Netherlands has a tradition of more localised shopping patterns than Britain. Planning policy has generally been more strict in attempting to control pressures for development. Pressure for out-of-centre retail development led to a revision of the Physical Planning Act in 1976 to make retail planning studies compulsory for development plans to investigate the existing retail structure of an area, consumer behaviour and future shopping prospects. But by the time these measures had been adopted, the boom in most kinds of peripheral retailing was over. The government reformulated its policy for large-scale retailing in 1984, but it remained restrictive towards out-of-centre development. The requirement for retail studies was lifted in 1985 (Borchert, 1988). Developers must now prove that development is not possible in sites on the edge of a town centre and is accessible by public transport before an out-of-town location will be considered. This policy is remarkably similar to the British government's policy as set out in PPG6.

As in Britain there is a growing concern in Sweden for the vitality and viability of town centres. Westlake and Forsburg point out that, unlike Britain, Sweden has no national policy towards retailing and local government is not required to produce retail policies or plans, although municipalities do have the power under the 1987 Planning and Building Act to influence the location of retail outlets. Traditionally politicians and planners have sought to defend the town centre as the main area for commercial activity in the municipality. Prior to 1992 municipalities were criticised by retailers for reinforcing the existing market structure by not granting permission for out-of-town centres or for siting food stores in industrial areas. Since 1992 municipalities have lost the power to restrict new retail development to residential or town centre locations.

> Sweden finds itself in the position of the UK in the mid-1980s with a steady decline of traditional town centres. If central and local government in Sweden do not respond quickly then it will be too late to preserve the vitality and viability of many town centres.
>
> (Westlake and Forsburg, 1996: 28)

Evidence of impact

There is very little hard evidence of the impact of out-of-centre retail developments in continental Europe. Initially it was assumed that the hypermarket would decimate the small shop sector in France, causing the widespread closure of outlets. But in practice hypermarkets appear to have had little impact on small independent traders. Supermarkets and competing multiples have tended to be more affected. An analysis of trends in hypermarket development by Burt (1985) suggests that the Loi Royer has had little impact upon the rate and type of new retail development.

However, more recent evidence is available from France and Germany. Investigations by the House of Commons Select Committee on the Environment (House of Commons, 1994) found that in France out-of-town developments have had a severe impact on town centres, particularly on the smaller towns. Officials of towns who were visited by members of the committee tended to place responsibility for these impacts on the mayors of neighbouring towns who took the planning decisions. In contrast, in the German town of Freiburg, the Environment Committee found that a highly planned, even authoritarian, approach had been adopted. The town plan focused some forms of retailing (food, flowers, clothes, shoes and textiles) in the town centre while allowing out-of-town development for bulkier items such as DIY goods, furniture and carpets.

Approaches to retail impact assessment

In France, under the Loi Royer, all proposals for major shopping development are considered under three broad criteria:

- national levels and trends of retail and commercial activity
- the evolution and structure of retailing in the *département*
- the balance between the types of trade in the *département*.

The planning commissions are provided with the results of an impact assessment study. Modifications to the Loi Royer in 1996 give an increased role to the commissions, as noted earlier, which include reducing the threshold on floor areas, and holding a public inquiry into proposals for large-scale retail developments to assess their social, economic and land use impacts.

The modifications also set out a number of general principles which the commissions should take into account in reaching decisions, such as estimates of demand and supply in the relevant retail catchment areas, assessments of existing provision of medium and large retail outlets, and the potential effects of the proposals on small traders. All new developments and substantial extensions to existing premises should be considered in the light of their impact on town centres, areas of urban regeneration and the environment. These changes have been the subject of considerable criticism. In the view of many small traders, the system will be insufficient to protect their position. In contrast, the large retailers advocate complete deregulation of retailing (Begg-Saffar and Begg, 1996).

Under the German planning regulations proposals for major retail development must satisfy certain criteria, including:

- infrastructure problems, e.g. traffic generation
- effects on the provision of convenience shopping for the local population
- effects on the urban structure, e.g. trade in other shopping centres
- amenity factors.

Experience shows that the possible effects depend on the location, type of shop, and size of development (TEST, 1989).

North America

Retail development in North America

Out-of-town shopping centres have had a much longer history in North America than in Europe. Their development can be traced back to the arcades of the nineteenth century. The 1950s saw the first fully enclosed, air-conditioned shopping malls which provided a new standard of physical environment for the shopper. Several factors led to the growth of out-of-town retailing in North America, notably suburbanisation, car ownership and increasing wealth. There were two major trends: decentralisation – the shift of retail activity from the central business district (CBD) to suburban areas, and the development and rapid growth of planned shopping centres.

This section focuses on the USA but also makes reference to Canada which has also experienced the growth of shopping malls.

Demographic factors have probably been the major influence on the decentralisation of retailing in the USA and the decline of the city centre. Decentralisation of population has been combined with higher rates of growth of consumer spending in the suburbs. Retail investment followed population dispersion in a relatively free market with few land and planning restrictions (Distributive Trades EDC, 1988). The key factor in retail decentralisation became the decisions of department store companies to develop 'junior' department stores in the new suburban centres and close their central area stores after several suburban stores had successfully been established (Guy, 1994a).

By 1990 there were 36,650 planned shopping centres in the USA, accounting for 55 per cent of retail sales (Goss, 1993). Almost 1,800 of these centres were of over 400,000 square feet and almost all are in suburban locations (Guy, 1994a). The Mall of America near Minneapolis is the largest shopping centre in the USA with more than 400 stores. As one writer puts it, 'suburbia has been malled' (Lord, in Dawson and Lord, 1985).

Planned shopping centres have evolved into four major types:

- *super-regional centres* – enclosed with over 100,000 square metres gross floorspace, over 100 shop units and several department stores
- *regional shopping centres* – usually enclosed with 40 to 100,000 square metres gross floorspace and several anchor stores
- *community shopping centres* of 10 to 40,000 square metres anchored by a discount department store
- *neighbourhood shopping centres* of up to 10,000 square metres anchored by a supermarket and drug store.

From the mid-1970s onwards these 'conventional' formats have been augmented by three new types of centre:

- *'theme' centres* composed of related specialty stores, e.g. selling fashion goods, mostly in city centres or affluent suburbs
- *'multi-use' centres* in healthier city centres, e.g. Boston, including significant office and entertainment provision as well as retail space
- *factory outlet centres* offering merchandise at discount prices (Rogers, in Davies and Rogers, 1984).

Since the 1970s there has been increasing specialisation of American retailing, with the rise of the specialty store at the expense of the department and variety store. A new trend has emerged with the growth of 'category killers' which couple wide selections and low prices in order to dominate their sectors. This reflects an increasing focus on 'everyday low prices' as a competitive tactic (Rogers, 1991).

The decline of the central business district (CBD) as the prime retail focus has brought about the development of downtown malls to attract shoppers back to city centres (Fairbairn, in Davies and Rogers, 1984). Although a number of American cities realised that retail revitalisation was essential, these mall developments in city centres have tended to be unsuccessful because the concept does not work well in an urban setting (Carey, 1988).

Canada experienced a rapid growth of purpose-built shopping centres in the 1960s. As well as the general growth in consumer demand, three factors combined to produce a new dimension to consumer demand:

- the rapid growth in the urban population, especially in Montreal, Toronto and Vancouver
- increased car ownership which allowed the development of extensive suburbs in many Canadian cities
- the activities of land investors and developers in developing planned shopping centres in suburban locations (Shaw, in Dawson and Lord, 1985).

West Edmonton Mall in Alberta, with 5 million square feet of floorspace, is the world's largest shopping centre development.

Since the mid-1980s the rate of development of new shopping malls in North America has declined substantially, mainly because the local markets in many areas have reached saturation point (Carlson, 1991). Competition between centres has remained strong in the 1990s, and many centres have faced financial difficulties. Developers are now looking to build new stores and centres on freestanding sites along major traffic arteries (Ghosh and McLafferty, 1991).

The planning response

To understand the planning response to pressures for out-of-town retailing in North America, it is necessary to refer briefly to the governmental context in which planning operates in the USA and Canada. The USA is a federal state divided into 50 states. Each state has its own legislature and constitution based on the federal model. The 50 states are divided into counties and incorporated cities. The federal government generally does not get involved in land use planning. Planning legislation and institutions are left to the State and local levels of government. Planning has tended to be concerned with zoning and the use of particular areas of land. States have 'goals' which set the basis for locally prepared plans. Retail development can proceed if it is 'correct for the zone' but planning permits are required for building (Thomas, 1992, and Teitz, 1996). 'Public policy control aimed directly at commercial development of all types in the USA traditionally has been minimal' (Dawson and Lord, 1985: 9).

In Canada planning legislation affecting retailing is provincially organised and there are disparities between provinces. Alberta, home of West Edmonton Mall, has generally chosen a very *laissez-faire* attitude. Ontario, on the other hand, where the 1983 Planning Act closely resembles the system in England and Wales, exerts rather more control. Provincial initiatives have been taken to curb regional shopping centres and there are both federal and provincial initiatives offering positive aid to downtown traders (Hallsworth, 1990a).

Therefore, in much of North America there is no attempt at state or local government level to regulate the provision of shopping development. Land use planning exists only in the form of zoning ordinances which are intended to ensure that 'bad neighbour' effects of large scale development are minimised. In this context the competitive effects of new development upon older shopping centres is not generally considered a legitimate focus for government intervention. Significant differences, however, exist between southern and northern USA and between western and eastern Canada. In the latter areas some controls have been introduced by state or provincial governments following the effects of previous unregulated developments of regional shopping centres (Guy, 1994a).

One of the few examples of public policy intervention in the USA is the community conservation guidance policy (CCG) operated by the Carter administration between 1979 and 1981. It authorised under certain circumstances the preparation of an impact analysis for new retail development which involved federal action or funding. A community impact analysis was to be carried out to consider the positive and negative impacts of any large-scale retail development which might have an adverse impact on existing centres. If a negative impact was predicted, federal agencies were not to provide support for the development of the project. But the policy met with strong opposition from the retail industry which was antagonistic to the programme before it was even clearly formulated (Dawson and Lord, 1985). The policy was terminated by the Reagan government in 1981. Although in effect for only a few years, CCG did slow down the development of regional shopping centres.

A more recent example of the response to the effects of out-of-town retailing in the USA is in the form of positive initiatives to help town and city centres which have suffered decline as a result of out-of-centre developments. City centres still have a captive market of business employees and residents, and visitors to the central area. There is an increasing number of revitalisation schemes in central areas. Some cities have introduced business improvement districts (BIDs) which enable property owners to be locally levied for an additional contribution towards higher standards of environment, security and promotion, and for improving the facilities of the town centre (URBED, 1994).

The recognition of the problems arising from out-of-centre retail developments in Canada has led to new retail planning policies. For example, in Toronto large new retail developments were encouraged to locate in proximity to existing shops and to be complementary rather than competitive. Policy statements were adopted in the city's official plan in 1980, reflecting a shift of policy to one of positive planning for those retail areas suffering owing to the impact of new shopping centres, and limits were placed on the development of retail floorspace outside traditional shopping streets (Shaw, 1987).

Evidence of impact

In a short space of time the planned shopping centre has spread rapidly to towns and cities throughout the USA. There is increasing concern about the impact of large shopping mall developments on traditional downtown areas. The retail function of the CBD has suffered a serious decline (Lord, in Dawson and Lord, 1985). Decentralisation of retailing has been so extensive that most people have no need to go into the traditional downtown to shop. Many US downtowns are now given over to other functions but many have vacant and derelict plots and buildings. This is sometimes known as the 'doughnut effect' (URBED, 1994).

> It has taken less than three decades to destroy the heartbeat of many
> American cities. It will take a century to repair the damage and return
> those cities to a semblance of health. Not all cities will survive.
>
> (Carey, 1988: 47)

The experience of the impact of out-of-town developments in Canada has been different from the USA. Canadian downtowns have not declined in the same way as has happened in the USA. Enclosed malls have been built in city centres, usually anchored by major department stores and made up of national multiples. Regional shopping centres are very popular in Canada but city centres still have an important role to play (Gayler, 1989). However, in Toronto the growth of purpose-built shopping centres has had a significant impact on existing shopping areas, particularly in the older parts of the city, resulting in a decline of many traditional retail strips outside the CBD and in high vacancy rates. Public policy has moved to one of limitations on the growth and location of new shopping centres (Shaw, in Dawson and Lord, 1985).

There is more recent evidence from Canada of the impact of large new stores. Wal-Mart moved into Canada in 1994 and now has 30 per cent of the Canadian department store market. It has been reported that every community within an hour's drive of a Wal-Mart has felt the effect. 'Many Canadian retailers failed to adjust and went out of business, while others are barely hanging on. Survivors have been forced to improve customer service and to reduce prices' (Eade, 1999).

Approaches to retail impact assessment

Some US states have attempted to control the development of large retail developments through legislative planning powers. Vermont's Land Use and Development Law (Act 250) passed in 1970 requires that developers of schemes larger than a critical size (10 acres), such as a shopping centre, obtain a development permit from the state. The proposed project must satisfy several criteria to test whether there would be an adverse effect on the ability of the municipality to provide services for its population. In practice only a small percentage of permits have been refused (Dawson and Lord, 1985). In 1978 the Vermont District Environment Commission denied approval for the development of a regional shopping centre on the outskirts of Burlington. It was estimated that a considerable amount of retail sales would be diverted from the downtown and there would be a net loss of jobs in retailing. The public hearing proved to be a test case and other civic authorities began to take a tougher line. The decision was further evidence of growing government resistance towards outlying regional shopping centres (Davies, 1984).

This attitude of resistance to commercial pressures in the USA in the late 1970s led to the CCG policy to curb the future spread of suburban shopping developments and encourage more investment in city centres (noted earlier). A set of guidelines was laid down on how controls could be exerted through a stricter allocation of public money for the infrastructure and services necessary to support new developments (Davies, 1984). Under the CCG policy a total of 24 studies were completed, 15 of these for proposed regional shopping centres, mainly in the north east of the USA. A central question in the studies concerned the likely impact of the proposed shopping centre on existing retail facilities. This impact was measured in terms of retail sales diversion and the accompanying impact on sales tax and property tax revenues.

> Since the CCG studies were conducted by several different firms, there was little consistency in methodology used to assess economic impact. Furthermore, many of the studies were quite vague in the specific methods used in the impact assessment.
>
> (Dawson and Lord, 1985: 14)

Norris (1992) also comments that no methodological framework was used in the USA. He notes that there has been more concern with impact issues in Canada but very few impact studies have been carried out. Few examples of impact studies in the USA are readily available but it is clear that assessments are based mainly on economic factors and that use is still being made of gravity-based models. This approach is perhaps more appropriate in the USA than in Britain because the scale of the settlement pattern is such that there is less overlap between the catchment areas of towns and cities, and so shopping behaviour can be predicted more accurately than in Britain. Norris (1992) concluded that there are no alternatives to current practice which can be easily imported into Britain.

Relevance of international comparisons

There are some lessons to be learnt from the experience of retail development in North America, but Hall and Breheny (1987) point out that out-of-centre development has proceeded much further in the USA and Canada than in Britain. In Britain investors and retailers have continued to have confidence in established centres. Guy (1994a) elaborates on the growth of out-of-town shopping in North America with reference to regional shopping centres. He states that:

> The regional shopping centre is a North American phenomenon which was the logical outcome of the needs of shoppers and retailers in an unregulated development market. Department stores migrated from traditional town centres and their customers quickly came to prefer the climate-controlled, clean and safe environment of the shopping mall compared with the sometimes unsavoury environment within the CBD.
>
> (Guy, 1994a: 310)

By comparison, regional shopping centres have not become so established in Britain, and Guy notes that most regional centres in Britain have been built in locations which are quite different from those typical in North America. The similarities and differences between the British and North American experience, therefore, can be very instructive in understanding the retail development process in Britain.

It is also extremely relevant to compare Britain with western Europe. Guy (1994a) points out that in western Europe the town or city centre generally has a much stronger role than in North America. It is of historical importance and has a business and entertainment function as well as a shopping function. Out-of-town centres are seen in Europe as being for shopping only. New out-of-town centres have faced opposition from established interests, and those that have been built, notably in France, have not always been successful.

It is not just the nature of retail development which is relevant in international comparisons; the planning and political context is also a major influence. The relaxation of British restrictions on out-of-centre development in the 1980s (although short-lived) was in contrast to the experience of several European countries and parts of Canada, where controls over suburban retail development have been intensified as the implications of such development for existing centres became apparent (Guy, 1994b).

> So far as retail development is concerned, there are two important aspects which differentiate the various national or provincial systems of land use control. The first is the extent of flexibility and discretion embodied in the system. The British system is unusually flexible: national and local planning policies are generally vague and allow considerable discretion to planning officials. In contrast, planning systems in most other European countries and North America are more precise, based around the principle of zoning ordinances which set out precise rules of land use and building design for specific areas. The second relevant dimension of planning systems is the extent to which there is strategic control over retail development. Here, European systems generally impose such control (often through specific legislation rather than land use planning guidance as such) whereas most states and provinces in North America do not.
>
> (Guy, 1994b: 92)

The House of Commons Select Committee on the Environment report, 'Shopping centres and their future' (House of Commons, 1994) recognised that France and Germany have markedly different approaches to retail planning and development than those in Britain. Members of the Environment Committee visited France and Germany in connection with their inquiry but the report makes little reference to the experience of dealing with problems of retail planning in Europe and its conclusions and recommendations do not point to any lessons from retail development in Europe. This is significant in itself because it suggests that there are no easy answers to the problems of out-of-centre retailing in Britain, and those answers may not be found elsewhere.

The role of the institutions which influence retailing is also a relevant consideration.

> The geography of retailing can be comprehended only if the role of the institutions which channel the demand from shoppers and retailers into particular types of retail development is also understood. The strength of these institutions – financial and governmental – in Britain has had decisive effects on the location and type of new retailing and has led to a very different pattern of development from that which obtains in North America.
>
> (Guy, 1994a: 310)

Summary

International comparisons are extremely relevant to an understanding of new retail development in Britain. Planned shopping centres have only become widespread in North America since the 1950s and in Europe since the 1960s. As in Britain they have had a very significant effect on patterns of shopping. The pattern of development, and the planning response to the pressures for retail development, have been very different in Europe and North America to that in Britain.

New retail development in Europe has mostly been in the form of hypermarkets in France and smaller superstores in other countries. In North America there has been a massive growth of planned shopping centres or malls. The relatively small number number of out-of-town regional centres that have been built in Europe have generally been less successful than in the USA.

Europe has adopted increasingly strict planning powers to control new retail development, notably the Loi Royer in France and the BauNVO in Germany. These powers have been strengthened in recent years. Policy in The Netherlands is very similar to that in Britain. In the USA the public policy response to commercial pressures for retail development has been minimal. There have been only very limited attempts to control new development in the USA, but greater efforts have been made in Canada.

Evidence of the impact of out-of-town retail developments is very clear in the USA, where CBDs throughout the country have suffered serious decline from the loss of trade. The scale of the impact is lower in Canada but still significant. The rate of development of new shopping centres in North America has been slowing down in recent years. The levels of impact are generally less significant in Europe. There is evidence of the impact caused by hypermarkets in France but little evidence from other countries of the effects of new retail developments.

There is no clear 'preferred' approach to assessing retail impact in North America or Europe, and the methodology is much less developed than in Britain. Decisions on proposed developments in North America and Europe are based mostly on economic factors, as in Britain, but there appears to be a greater concern than exists in Britain about social and environmental impacts.

Evidence from Europe shows that there are few lessons to be learnt about the application of RIA in Britain, except for the way in which impact assessment is used as the basis of the implementation of the Loi Royer in France. The extent of decentralisation in North America and the scale of development of planned shopping centres are so unlike the situation in Britain that it is difficult to draw any meaningful comparisons. The lack of an effective planning system in the USA, and the lack of importance attached to impact assessments, is testimony to the merits of the planning policy framework that exists in Britain.

10

THE APPLICATION OF
BEST PRACTICE

Previous chapters have set out an approach to RIA which incorporates both quantitative and qualitative factors in the context of planning policy. The relevance of RIA as a tool for decision-making has been discussed. In this chapter attention is focused on the practical application of RIA. First, an illustration is given of the application of the quantitative impact assessment framework using a hypothetical example of a proposed superstore development, but it is closely based on a real situation which has been the subject of a retail impact study. A concise checklist is then presented for undertaking a RIA, which is recommended as a model for best practice.

Illustration of the impact assessment framework

In Tables 10.1 to 10.3 a simplified illustration is given of the application of the recommended framework for quantitative impact assessment using a hypothetical catchment area divided into sub-areas, and with a range of shopping centres. The illustration is based on the impact of a proposed superstore development, but it does not matter whether the data shown represents convenience or comparison expenditure; the principles are the same for both. In Table 10.1 base year expenditure flows are shown between sub-areas and centres, and in Table 10.2 expenditure is projected for the design year. Table 10.3 illustrates the impact matrix which is used to assess the impact of a proposed new development.

In this hypothetical example the retail context is an urban area which includes a town centre, a district centre, a local centre and other local shops. It is located near to a conurbation. The total population of the primary catchment area is about 80,000. There are five sub-areas. The scenario to be tested is a proposed food superstore development of 3,000 square metres net floorspace located on the edge of the town centre. The expenditure is defined on the convenience business basis and is in 1995 prices. The design year is set as one to two years after the likely opening of the store, and three years from the base year.

Table 10.1 Base year matrix, convenience businesses, expenditure (£000) in 1995 prices

Sub-areas	Proposed new store	Town centre				District centre	Local centre	Other shops	Leakage	Total expenditure
		Store A	Store B	Other shops	Total					
1	–	5,000	2,700	4,900	12,600	2,000	1,300	2,200	6,400	24,500
2	–	1,200	600	1,200	3,000	–	–	1,800	13,700	18,500
3	–	800	400	800	2,000	2,000	1,300	2,600	13,600	21,500
4	–	2,500	1,700	2,400	6,600	4,900	3,400	4,400	14,300	33,600
5	–	500	300	500	1,300	1,000	–	1,100	10,500	13,900
Catchment area total	–	10,000	5,700	9,800	25,500	9,900	6,000	12,100	58,500	112,000
Inflow of expenditure	–	2,000	1,000	500	3,500	3,000	–	–	total inflow	6,500
Total turnover	–	12,000	6,700	10,300	29,000	12,900	6,000	12,100	total turnover	60,000
Floorspace (sq. metres net)	–	1,800	1,200	2,800	5,800	2,800	1,800	3,300	net leakage per cent	52,000 / 46%
Turnover per sq. m. (£)	–	6,667	5,583	3,679	5,000	4,607	3,333	3,667	retention level	54%

Notes
Store A is a large supermarket. Store B is a smaller supermarket. The district centre contains a large supermarket. Leakage is relatively high because there are existing superstores within travelling distance. All figures are rounded.

Table 10.2 Design year matrix, convenience businesses, expenditure (£000) in 1995 prices

Sub-areas	Proposed new store	Town centre				District centre	Local centre	Other shops	Leakage	Total expenditure
		Store A	Store B	Other shops	Total					
1	–	5,300	2,900	5,200	13,400	2,100	1,400	2,300	6,900	26,100
2	–	1,300	600	1,300	3,200	–	–	1,900	14,500	19,600
3	–	800	400	800	2,000	2,100	1,400	2,800	14,500	22,800
4	–	2,700	1,800	2,500	7,000	5,200	3,600	4,700	15,200	35,700
5	–	500	300	500	1,300	1,100	–	1,200	11,200	14,800
Catchment area total	–	10,600	6,000	10,300	26,900	10,500	6,400	12,900	62,300	119,000
Inflow of expenditure	–	2,000	1,000	500	3,500	3,000	–	–	total inflow	6,500
Total turnover	–	12,600	7,000	10,800	30,400	13,500	6,400	12,900	total turnover	63,200
Floorspace (sq. metres net)	–	1,800	1,200	2,800	5,800	2,800	1,800	3,300	net leakage per cent	55,800 / 47%
Turnover per sq. m. (£)	–	7,000	5,833	3,857	5,241	4,821	3,556	3,909	retention level	53%

Notes
Expenditure flows have been increased pro rata to expenditure growth in each sub-area. Inflows remain unchanged. All figures are rounded.

Table 10.3 Impact of a new superstore on the edge of a town centre, convenience businesses, expenditure (£000) in 1995 prices

Sub-areas	Proposed new store turnover	Trade draw turnover	Town centre				District centre	Local centre	Other shops	Leakage	Total expenditure
			Store A	Store B	Other shops	Total					
1	30%	6,750	4,400	2,500	4,700	11,600	1,800	1,300	2,200	2,450	26,100
2	20%	4,500	1,100	500	1,200	2,800	—	—	1,800	10,500	19,600
3	10%	2,250	700	300	800	1,800	1,900	1,300	2,600	12,950	22,800
4	20%	4,500	2,500	1,700	2,400	6,600	4,700	3,400	4,500	12,000	35,700
5	10%	2,250	400	300	500	1,200	1,000	—	1,100	9,250	14,800
Catchment area total	90%	20,250	9,100	5,300	9,600	24,000	9,400	6,000	12,200	47,150	119,000
Inflow of expenditure	10%	2,250	2,000	1,000	500	3,500	3,000	—	—	total inflow	8,750
Total turnover	100%	22,500	11,100	6,300	10,100	27,500	12,400	6,000	12,200	total turnover	80,600
Pre-impact turnover	—	—	12,600	7,000	10,800	30,400	13,500	6,400	12,900	total turnover pre-impact	63,200
Trade diversion	—	—	1,500	700	700	2,900	1,100	400	700	total trade diversion	5,100
Percentage impact	—	—	12%	10%	6%	10%	8%	6%	5%	net leakage % of expenditure	38,400 / 32%
Floorspace (sq. metres net)	—	3,000	1,800	1,200	2,800	5,800	2,800	1,800	3,300	existing gross leakage clawback	58,500 / 11,350
Turnover per sq. m. (£)	—	7,500	6,167	5,250	3,607	4,741	4,429	3,333	3,697	% of existing gross leakage % of store turnover	19% / 50%

The base year matrix (Table 10.1) has been set up using estimates of turnover based on market shares from a household survey. It shows a significant gross leakage from the primary catchment area to superstores in the nearby conurbation. There is a small inflow of expenditure into the area from beyond the primary catchment area boundary, which includes spending by visitors. The net leakage is 46 per cent of total convenience expenditure. Within the town centre there is a large supermarket (Store A) and a smaller supermarket (Store B). The district centre also has a large supermarket. The town centre has a market share of about 23 per cent of convenience spending by residents.

In the design year matrix (Table 10.2) expenditure flows are increased pro rata to expenditure growth in each sub-area. Expenditure projections are made using a growth rate of 2.0 per cent per annum for convenience businesses and taking account of population growth. Leakage also increases in line with expenditure growth. For simplicity, it is assumed that inflows remain unchanged from the base year. Some consultants prefer to keep turnovers of existing stores and centres at their base year level in the design year because it cannot be assumed that all stores and centres will maintain their market share in the future. Increased expenditure in the design year is then reflected in an increased amount of gross leakage, which is unrealistic. It is also unrealistic to expect that no growth in turnover will occur between the base year and design year. Turnovers are the sum of expenditure flows to particular centres. These turnovers represent pre-impact turnovers as the basis for assessing trade diversion post-opening.

In the impact matrix (Table 10.3) the turnover of the proposed store is £22.5 million, calculated using a typical turnover/floorspace ratio of £7,500 per square metre for this type of store. The matrix shows the estimated percentage trade draws from each sub-area and inflow. In the impact matrix adjustments are made to the design year expenditure flows to subtract the trade drawn to the new store from the total convenience expenditure generated. The remaining expenditure is then available to be spent in existing stores and centres. The allocation of this residual expenditure between stores/ centres must be based on subjective judgement. It has to replicate the way that shoppers would change their shopping behaviour when a new store opens. New superstores tend to compete with other large foodstores like-with-like, rather than take trade from local shops. The expenditure flows after trade diversion to the new store has been estimated are aggregated to give post-opening turnovers of each existing store/centre. Inflows remain the same but the total inflow is increased because of inflow to the new store. The difference between total turnover and pre-impact turnover (from the design year matrix) is total trade diversion, which is expressed as a percentage of the pre-impact turnover. Impact is also shown in the matrix as residual turnover per square metre of each store/centre.

The turnover of a new retail development will be made up of a number of components. These include trade diversion from existing centres and clawback of leakage, but also the effect of increased inflow to the new development and an element of expenditure growth up to the design year. The components of the turnover of a new store are shown in the example given in Table 10.4.

In Table 10.3 net leakage falls from 46 per cent to 32 per cent of expenditure. The amount of clawback is 19 per cent of existing gross leakage. Clawback is a more significant factor in this instance than trade diversion as it provides half of the store's turnover. This amount of clawback appears high but there is evidence that shoppers who are already using out-of-centre superstores some distance away are willing to transfer their trade to a closer store. This level of clawback would also represent a significant reduction in the need to travel for main food shopping. The new store is predicted to take 54 per cent of expenditure growth in the primary catchment area, showing that existing stores will also benefit from the growth in available spending.

It is preferable to express clawback in relation to existing leakage rather than estimated leakage in the design year. All centres are assumed to benefit pro rata from the increase in expenditure flows which take place between the base year and the design year. However, it is unrealistic to expect that all shops will share in the growth of expenditure to the same extent. The increase in turnover will, in reality, be influenced by the competitiveness of shops in the catchment area compared to those elsewhere. The components of turnover shown in the example above include both clawback of (existing) leakage and expenditure growth. Any new retail development will take a share of expenditure growth from its opening to the design year. The growth of expenditure is particularly important in the case of comparison goods where a significant proportion of the turnover of a new development may come from expenditure growth in the catchment area. In catchment areas with tourist spending it is also necessary to take account of expenditure by visitors to the area.

Interpretation of the impacts of the proposal must take account of the vitality and viability of centres in the area. In this illustration it is assumed that all the centres in the area are trading at viable levels in the base year. In

Table 10.4 Components of the turnover of a new store

	£000	per cent
Trade diversion	5,100	23
Reduction in leakage (clawback)	11,350	50
Increased inflow of trade	2,250	10
Expenditure growth from base year	3,800	17
Total	*22,500*	*100*

Table 10.5 Linked spending

	Linked spending per £1 spent at Somerfield
In-town stores	46p
Edge-of-centre locations	21p
Out-of-centre stores	10p
District and local centres	15p

the town centre the two supermarkets (Stores A and B), have predicted trade diversions of 12 per cent and 10 per cent respectively. These levels of impact are judged to be not significant in the light of the residual turnovers which remain at viable levels. Other shops in the town centre would not be affected significantly, with a 6 per cent trade diversion. The overall impact on town centre convenience trade is 10 per cent. Some trade diversion is inevitable when a new store opens and changes shopping patterns. In a town centre with an above average level of vitality and viability, an impact of this order is not likely to be significant, as shown earlier in Figure 6.3. Predicted impacts, in terms of trade diversion and residual turnover, are also judged to be not significant in the district centre, in the local centre or in other local shops.

Used as the basis for assessing quantitative impact, the application of the impact assessment framework in this illustration shows that the 'impact test' would be satisfied and the proposed development would not harm the vitality and viability of any existing centres.

In assessing the impact of a proposed superstore development it is necessary to take account of the mitigating effects of spin-off benefits from linked trips. In 1977 Somerfield Stores commissioned an interview survey of shoppers to identify the extent of linked trips to 33 of its stores in a range of locations. The survey results indicate that linked shopping is a very significant feature of retail spending in many centres. The average amount of linked expenditure varied between different categories of centre, as shown in Table 10.5 (Roger Tym and Partners, 1997).

The biggest beneficiaries of linked spending are chemists, newsagents, smaller food shops and clothing/footwear shops. The results show that a well located foodstore can provide linked spending benefits for town centre retailers. The implications of these findings are that trade diversion from a town centre arising from the opening of a new foodstore can be offset by the spin-off benefits resulting from linked shopping in the centre. This factor is not built into the quantitative assessment but it should form part of the qualitative assessment of retail impact in the checklist shown later.

If this recommended framework is used to test the impact of a proposed retail park or a factory outlet centre, instead of a superstore, the principles would remain the same but the following points should be noted:

- the impact of comparison shopping developments is normally assessed using expenditure on a goods basis; the net floorspace in stores which sell a mix of convenience and comparison goods must be classified into convenience and comparison floorspace
- the catchment area for comparison shopping will be larger than for convenience shopping, particularly in the case of a factory outlet centre; this difference should be reflected in the study area used for the household survey and in the definition of the primary catchment area
- in the design year matrix the amount of expenditure growth will be much higher for comparison goods than for convenience goods; therefore the design year turnovers will also show a significant growth between the base year and the design year. However, it would be unrealistic to expect all expenditure growth to be allocated between existing centres. In the impact matrix the pre-impact turnovers should be derived from the base year figures increased at a rate of 1.5 to 2.0 per cent per annum to reflect the growth of floorspace efficiency for comparison goods shops
- in interpreting the impact of a proposed comparison shopping development, the significance of the predicted trade diversions should be assessed in the context of the growth of comparison goods expenditure; an 8 per cent trade diversion would be offset in two years by expenditure growth.

A checklist for retail impact assessment

Previous chapters have recommended advice on best practice in RIA. This advice is now brought together in the form of a checklist for the practical application of RIA. It includes all the qualitative elements required by PPG6 as well as the quantitative elements that were illustrated in the impact assessment framework earlier in this chapter. The checklist embodies the advice given in DETR/CB Hillier Parker's CREATE approach shown in Figure 5.1 but it sets out the steps involved in the approach in more detail. The recommended format can be used as an outline of the content of an RIA report, but of course the level of detail will vary according to the complexity of the particular case. The checklist is not concerned with employment impact or with environmental impact, both of which require a specialist methodology which is beyond the scope of this book. Employment analyses are not normally carried out as part of RIAs. Environmental impact assessments may be required for very large retail schemes but an EIA will be a separate exercise from the RIA.

1 Introduction
The context for the RIA
Outline of approach used

2 The proposed development
Location
Type of shopping – convenience, comparison, mixed
Floorspace – gross and net
Employment

3 Shopping policy context
National policy guidance
Regional guidance
Development plan – relevant policies

4 Definition of the study area
Isochrones – identify 5-, 10- and 15-minute off-peak drive time bands from the proposal site
Catchment area and sub-areas – define primary catchment area and divide into sub-areas

5 Review of existing shopping provision
Shopping centres – hierarchy of centres
Recent shopping developments
Floorspace – existing floorspace by sector

6 Health check appraisal
Health check indicators and factors (see Figure 6.1):
- diversity of uses
- retailer representation
- vacant properties
- commercial performance
- pedestrian flows
- accessibility
- customer views and behaviour
- safety and security
- environmental quality

Vitality and viability index

7 Shopping patterns
Household interview survey – sample size, questionnaire design
Analysis of shopping patterns
Linked trips

8 Need and capacity
Parameters – define base year, design year and price base
Existing retail expenditure – population and per capita expenditure on goods or business basis
Existing turnover estimates – market shares based on household survey

Leakage and retention level

Expenditure projections:
- population and per capita expenditure
- choice of growth rate of expenditure

Capacity analysis:
- expenditure growth
- potential for clawback of leakage
- committed developments
- floorspace capacity

Evidence of quantitative need:
- economic capacity/demand
- leakage of trade
- retailer requirements

Qualitative need:
- any deficiencies in provision
- indications of over-trading
- consumer demand

9 Sequential site assessment

Identification of alternative sites

Evaluation of sites:
- suitability
- viability
- availability

10 Quantitative impact assessment

Expenditure flows in base year (see base year matrix in Table 10.1) – disaggregate population and expenditure by sub-area

Turnover of proposed development – company averages adjusted to local trading conditions

Trade draw – estimate percentage trade draw from sub-areas and inflow

Design year expenditure flows (see design year matrix in Figure 10.2)

Impact analysis (see impact matrix in Figure 10.3)

Cumulative impact (if applicable)

Trade diversions:
- amount and percentage
- subjective judgement

Spin-off benefits through linked trips

Residual turnovers per square metre

Components of turnover of store:
- trade diversion
- clawback
- increased inflow
- expenditure growth

Sensitivity analysis:
- level and growth of per capita expenditure
- predicted turnover of proposed development
- amount of clawback

Interpretation of economic impact:
- model of significance of impacts
- ability of centres to withstand impact
- risk of closures of shops

11 Qualitative assessment of impact
Accessibility by a choice of means of travel:
- existing bus services
- accessibility by bus and on foot/cycle
- potential for improvements in public transport/pedestrian links
- potential for linked trips

Travel and car use:
- characteristics of catchment population
- proportion of car-borne trade
- potential for clawback of leakage
- reduction in length of car journeys

PPG6 paragraph 4.3 implications:
- risk to town centre strategy
- effect on future investment
- changes to quality, attractiveness and role of centre
- changes to physical condition of centre
- effect on range of services
- increases in vacancies

Other material considerations

12 Conclusions
Evidence of need – quantitative and qualitative
Sequential approach
Consistency with PPG6 guidance
Harm to development plan strategy
Impact
Sustainability
Benefits of the proposal

This checklist is consistent with the guidelines for RIA in PPG6. These guidelines are very general and say simply that:

- impact must be judged on the vitality and viability of existing centres – using health checks to assess vitality and viability

- a broad approach is recommended to the assessment of impact; parties should, where possible, agree data to be used and present information on areas of dispute in a succinct and comparable form
- a long-term view should be taken of retail impact; the full impact of the development may take some time to be felt
- the validity of any assessment will depend particularly on the quality, quantity and timeliness of the retail surveys undertaken (as specified in Annex B of PPG6).

Health checks are part of the framework set out above. A broad approach is sensible, as is agreement between parties on data and assumptions, particularly where RIAs are being prepared as evidence for an inquiry. Excessive detail should be avoided. Experience shows that a concise approach is more acceptable to local authorities and planning inspectors than the production of complex technical reports which may be incomprehensible.

The checklist approach in this chapter is designed to produce concise and comprehensive RIA reports. One of the difficulties for consultants acting for developers or retailers promoting a new retail development is that local authorities can sometimes be over-zealous in their requirements for information, demanding a level of detail of information which is unreasonable or which is not justified because of the uncertainty about the inherent quality and accuracy of the data itself.

Taking a long-term view is important – it may be appropriate to select a design year as much as five years ahead because of the time it takes for shopping patterns to adjust to the opening of a new store or centre.

The final point is to emphasise the need for the intelligent use of data. Acceptable assessments cannot be prepared using poor data. The use of the RIA checklist in this chapter needs to be supported by good surveys and analyses of the amount and quality of shopping provision, the vitality and viability of centres, and existing shopping patterns.

11

CONCLUSIONS

Summaries have been included at the end of the main chapters in this book. The conclusions in this chapter do not repeat the main findings but are presented in the form of a number of issues which are considered under the following headings:

- approaches to RIA
- the relationship between impact and policy
- implications of government policy on retail planning
- advice on best practice
- other lessons to be learnt.

Approaches to retail impact assessment

The evolution of RIA moved from the use of shopping models in the 1970s to the development of the step-by-step approach, which remains the most commonly used method of assessing retail impact. The use of models has almost ceased as problems of data inputs and calibration became insurmountable, and as confidence in their results declined. The step-by-step approach is more readily understood and can be applied without the need for specialist computer software. Consultants and researchers have attempted to refine the approach but problems of data and assumptions about the key variables limit the reliability of the conventional methodology of quantitative impact assessment. Minor variations in the assumptions can produce wide disparities in the analysis of impact. Quantitative assessment must be combined with a high degree of professional judgement so that decision-makers can have confidence in the accuracy of the results. The advice on best practice in this book highlights a number of ways in which the conventional methodology can be improved.

The recommended approach to best practice is that the application of RIA should focus on economic impact through an improved methodology for quantitative impact assessment, but that this should be combined with a thorough qualitative assessment of the effects of a proposed development

199

which includes the environmental impacts prescribed in PPG6, including the 'sustainability tests' of accessibility by a choice of means of transport and impact on overall travel and car use.

The emphasis in this book has been on economic impact because retailing is an economic activity and it is necessary to assess impact principally in terms of the vitality and viability of shopping centres and the effects of changes in shopping patterns on the trading position of centres. However, PPG6 advice has broadened the range of factors to be considered, and there is an increasing concern with environmental impact in terms of the sustainability of retail development. PPG6 is a direct response to the commitment of the government to sustainable development. The environmental principles of PPG6 and PPG13 are transport-based in seeking to reduce the use of the car for shopping trips and ensuring that new retail developments are in locations that can be served by public transport as well as by car. The effectiveness of this policy in controlling future retail development is uncertain. In an age of increasing car ownership, how can retailers ignore the preference of the majority of car-owning shoppers to use their cars for shopping trips, and how can shoppers be persuaded to leave their cars at home and use public transport? Unless major progress is made towards improvements in public transport, there are no real prospects of a major change in consumer behaviour in the foreseeable future.

Relationship between impact and policy

Retail planning policy has evolved in parallel with the development of planning theory over the past 30 years. The policy context is closely interrelated with planning theory. From the traditional origins of physical planning up to the 1960s, the concept of rational planning to guide urban and regional change led to a systems approach and the rational decision model which dominated the planning process in the 1970s. In the 1980s free-market ideology became prominent, influencing government policy on planning, leading to deregulation of retailing and a *laissez-faire* attitude to development. By the late 1980s, however, a more pragmatic approach was being adopted and it was accepted that it was legitimate for government to control the retail market through the planning process. Issues of retail impact imply a degree of government intervention over the operation of the retail market. In the 1990s planning has taken a more pragmatic approach, and government policy seeks to regulate the location of new retail development. It restrains free-market forces and introduces environmental as well as economic factors into decisions on major retail development.

The relationship between the development of planning theory, retail planning policy and RIA is illustrated in Table 11.1.

Several versions of PPG6 have shifted the emphasis towards a policy on retail development which recognises the need to support town centres and

Table 11.1 Evolution of planning theory, retail planning policy and retail impact assessment

	1960s	*1970s*	*1980s*	*1990s*
Planning theory	Planning as physical development	Rational planning model	Neo-Marxism and free market ideology	Pragmatism
Retail planning policy	Hierarchy of shopping centres	First efforts to control the development of large new stores	*Laissez-faire* policy	Emphasis on vitality and viability of town centres
Retail impact assessment	Shopping models	Post-hoc studies	Predictive assessments	Practical approaches

the need for a plan-led approach to the location of new development. Sustainability factors have also become much more significant. Policy and practice in Scotland and Wales are generally consistent with the guidance in England. The plan-led approach has been established as a policy mechanism in planning legislation and PPG6. The sequential approach has also become an important consideration by inspectors in assessing proposals for retail development.

Through successive revisions of PPG6, government policy on retailing now seeks to judge the effects of new retail development on the vitality and viability of shopping centres. Vitality and viability is seen as a reflection of the health of town centres. A health check approach to assessing vitality and viability is recommended in PPG6, using qualitative indicators to make judgements about vitality and viability. Although a range of indicators is suggested, there are conflicting views of the relative importance of different indicators. Commercial indicators such as yield are particularly difficult to interpret and their practical value is questionable. The best practice advice in this book is to use an appraisal framework which enables a vitality and viability index to be calculated from a wide range of indicators and factors. The index has been shown to be an accurate and useful method of making a qualitative assessment of town centres, and making comparisons between centres.

Changes in retailing have occurred faster than the planning system is able to respond. The government's response to decentralisation of retailing in the form of superstores, retail parks, regional shopping centres and factory outlet centres, has been to safeguard town centres. The government recognised that some town centres were in decline, partly owing to the effects of out-of-centre retailing and partly because town centres have failed to maintain their attraction to shoppers. Government policy now aims to promote, rather than just protect, the vitality and viability of centres. Making town centres more attractive to shoppers needs to take account of consumer demand. Out-of-centre developments are popular with shoppers, and shoppers will visit the centres they like best. Therefore town centres must meet consumer demand in competing with new forms of retailing. It is generally accepted that the balance in shopping policy had shifted too far from town centres to out-of-centre facilities before the latest revision of PPG. This balance needs to be restored through positive action on promotion and investment in town centre facilities.

RIA developed out of the new ideas of procedural planning theory and social engineering in the 1960s, but approaches to RIA have become less theoretical and more pragmatic in response to the requirements of the planning system. Impact assessment was broadened by the 1993 version of PPG6 and the emphasis has changed again in the June 1996 revision. The key advice now is the 'impact test' combined with qualitative assessments of town centres and other factors based on the sustainability of development. The requirement for quantitative assessment remains but there is much more emphasis on qualitative assessment. However, the latest guidance is still relatively new. Even in the early 1990s RIA was still seen as a largely quantitative process. The criticisms of local authorities about retail impact studies revealed by the survey in this research show that from a local authority viewpoint the overall quality of RIAs needs to be improved. This is backed up by the views of planning inspectors. Approaches to RIA have not kept pace with changes in the policy context in recent years.

Despite its shortcomings, RIA is still relevant because it is the means of implementing policy guidance. RIA is the basis for assessing the potential impacts of proposed developments in the context of the current policy guidance. The need to assess impact is perhaps most evident in planning appeals. Local authorities believed that too many out-of-centre developments were allowed on appeal during the 1980s and early 1990s, and a large number of proposals are still decided by inspectors through appeals or call-ins. The most important factor in appeals decided in Great Britain since 1988 has consistently been evidence of the lack of impact on nearby centres. Impact has also been the most important reason for dismissals. It is crucial, therefore, that decisions are soundly based on good planning evidence of which retail impact is still the paramount factor. RIA must be able to provide the basis for good advice on retail impact issues.

Implications of government policy on retail planning

Retail development is demand-led and it requires the planning system to control the location of new development. PPG6 is of fundamental importance in the control process because it lays down advice on the formulation of shopping policy in development plans and it gives guidance on how to assess proposals for retail development. The latest guidance is a marked shift from the policy of the 1980s which allowed considerable freedom in the location of new development and which resulted in the rapid spread of out-of-centre superstores, retail parks and regional shopping centres. The major change of emphasis in policy has been towards the positive support for town centres which represents a recognition of the need to reverse the decline of town centres.

The most significant change in policy, however, is the shift towards sustainable development. The 1990s can be regarded as the decade of sustainable development and the planning system has had a central role to play in implementing policy on sustainability. The environmental principles underlying sustainable development have been embodied in government policy but the practical application of the concept in the planning system is more problematic. In the retailing context, government policy seeks to achieve sustainability through locating new development where it is accessible by a choice of means of transport and by reducing travel demand. Emphasis is now being placed by inspectors on the PPG6 'sustainability tests'. A proposal may satisfy the 'impact test' but fail on the grounds of accessibility or level of car usage. The objectives of sustainability are sound but the problem is that at present, for most shoppers, public transport is not an acceptable alternative to the car. Pressure for further out-of-centre development seems likely to continue, but with developers and retailers looking for ways to make their proposals more acceptable in meeting the sustainability criteria. There is also a considerable amount of out-of-centre retail development in the pipeline from past planning consents which have not yet been implemented.

Sustainability objectives imply a different perspective on future retail development. To take a strategic view it is necessary to look beyond the development plan 10-year timescale to predict what the future pattern of shopping might be. To what extent, for instance, will town centres be able to maintain or strengthen their role for major shopping? Will any further regional shopping centres be allowed and how will they affect shopping patterns? Will it be possible to encourage shoppers to use public transport rather than cars? What will be the implications of emerging trends in tele-shopping? These questions really mean that there needs to be much more stability and consistency in government policy towards retailing than has been the case in the last decade.

The policy guidance in PPG6 has been well received by local authorities but it remains to be seen how the development industry will respond to the more restrictive stance on the location of new development. The sequential approach is particularly demanding on developers and retailers because it requires greater flexibility in site requirements. The major food retailers tend to be the most rigid in site location and layout, and it may not be realistic to expect to find sites capable of accommodating new superstores or large supermarkets in or on the edge of town centres. A trend towards smaller foodstores may be anticipated to meet the PPG6 site criteria, and avoid the prospect of many proposals for out-of-centre superstores being refused or dismissed on appeal because they fall foul of the key tests.

The other area in which the new guidance has a fundamental bearing on future retail development is advice on development plans. Development plans must provide a clear policy base for new shopping development. Policies must be able to deal effectively with pressures for development and so they must be up to date. Local plans in particular must also contain an up-to-date analysis of need and the likely impact of any proposed site allocations for new retail development. Development plans have taken account of changes in government policy guidance as they have been modified or reviewed, but in some cases policy lags behind government guidance and needs to be revised further to make it consistent with the latest guidance, and it will take time for this to happen.

Many factors influence whether a new retail development will have a significant impact, such as the size and type of development and the strengths and weaknesses of the centres affected. Most importantly, the predicted level of impact must be interpreted in the light of the vitality and viability of particular centres. Deciding what is a significant level of impact is very difficult. Trade diversions of less than 10 per cent have been considered unacceptable by local authorities and inspectors in some situations and impacts of more than 20 per cent have been regarded as acceptable in other cases. Relatively low trade diversions can be considered significant if they would harm the vitality and viability of particular centres. Cumulative impact has become a key factor in decisions on large-scale food and non-food developments. The effects of trade diversion are not always clear. It does not follow that a reduction in trade necessarily results in a loss of profit. Shops can adapt by economising, by becoming more efficient or by seeking rent and rates reductions.

There is a growing body of evidence about the effects of out-of-centre retail developments in Britain. Most of the evidence relates to food superstores, but there is now a substantial amount of evidence about the impact of retail warehouses/parks and out-of-town regional shopping centres. Retail warehouses and parks have tended in the past to sell bulky goods which could not readily be accommodated in town centres but there is a trend towards the sale of comparison goods which poses more of a risk of loss of trade from town centres. The impact of large retail parks can be significant but decisions are usually made on qualitative rather than quantitative factors, e.g. reducing the prospects for future investment in town centres. Out-of-town regional shopping centres can have a significant impact on nearby town and city centres, as the evidence of Meadowhall and Merry Hill shows. PPG6 recognises the serious implications of allowing any further regional shopping centres or out-of-centre factory outlet centres.

Advice on best practice

Government guidance on retail planning and the assessment of retail impact
is set out in PPG6. There is no other source of official advice on how impact
should be assessed. The House of Commons Select Committee on the Environment called for clearer and more detailed retail planning guidance for
inspectors and local authorities, in the form of a handbook, but this advice
was not accepted by the government. Instead, PPG6 gives only very general
guidance on assessing retail impact. Impact issues often lead to conflicts
between the parties at public inquiries over the factors involved in quantitative assessments. This situation is not helped by the paucity of official advice
on how retail impact should be assessed.

PPG6 contains advice on a number of aspects of retail impact:

- There is a requirement to assess future need or capacity for further retail
  development in an area as part of the sequential approach. The government has attempted to clarify the definition of 'need' and it is now
  necessary to examine a range of quantitative and qualitative factors according to local circumstances.
- A broad approach is recommended to the assessment of impact. PPG6
  implies that it is not necessary to undertake detailed quantitative analysis of the type which has led to criticisms of RIA in the past. However,
  a broad approach should not be interpreted as meaning a superficial
  approach. RIA must be systematic and soundly-based.
- A long-term view should be taken of retail impact. It is appropriate
  to look ahead several years in assessing impact because of the time taken
  for the effects on shopping patterns to become apparent.
- The emphasis on the vitality and viability of town centres means that
  impact must be judged on the health of centres. PPG6 includes a range
  of indicators of vitality and viability but there is no advice on how they
  should be used in practice. The best practice recommendations in this
  book show how it is possible to devise a vitality and viability index for
  a shopping centre which goes further than the PPG6 guidance.
- The cumulative impact of retail developments needs to be assessed. The
  assessment of cumulative impact should include recent developments
  that have taken place in the previous five years, as well as proposed
  developments which have planning permission and must be regarded as
  commitments.

This is as far as government advice goes. It does not indicate what is best
practice or give any advice on the technical issues involved in RIA. The
current guidance in PPG6 is not adequate for practitioners in the field.
There are still inherent problems with the application of RIA in Britain,
and there is a need for improvement in both quantitative and qualitative

assessment. It has been shown that the credibility of the application of RIA depends very much on the quality of the data and the assumptions used. The major criticism of lack of objectivity in retail impact studies can be overcome to some extent by carrying out independent audits or reviews of RIAs, and many local authorities now commission such reviews, but there is a clear need for better guidance.

The advice on best practice in this book makes a number of recommendations on the application of quantitative impact assessment, as follows:

- the careful definition of the catchment area of a proposed development and sub-areas which reflect the spatial characteristics of the retail system
- the use of household interview surveys to obtain an accurate representation of existing shopping patterns between sub-areas and centres/stores
- the estimation of the turnover of centres and large stores using market shares based on the household survey
- examining evidence of quantitative need in terms of economic capacity, leakage of trade and retailer requirements
- assessing quantitative impact on the basis of expenditure flows between sub-areas and centres/stores
- estimating trade draw to a proposed development according to population distribution in the catchment area and the scale and location of competing shopping facilities
- applying professional judgement in assessing trade diversion from existing centres, taking account of the nature of competition in the local area (on the basis that retail developments tend to compete 'like with like')
- expressing impact in terms of percentage trade diversion and residual turnover per square metre, comparing the predicted turnovers of centres with the minimum levels that can be considered viable for different types of centre
- testing the sensitivity of assumptions to build in an allowance for variations in assumptions and errors in forecasting, particularly in terms of the future level of expenditure, the predicted turnover of the proposed development, and the amount of clawback of leakage.

A comprehensive framework for RIA has been developed and refined in the course of the research for this book. It is a very practical approach to predicting the impact of a proposed retail development on future shopping behaviour. The framework concentrates on economic impact and involves constructing a matrix of expenditure flows for a retail sector in the base year. Expenditure from sub-areas is allocated between centres based on market shares derived from the household survey. The matrix forms the basis for predicting the retail impact of a proposed new development. A design year matrix is set up, projecting expenditure forward to the design year. In the impact matrix a proposed new store is introduced into the design year matrix and its likely impact assessed on expenditure flows. This framework is a marked improvement on the conventional step-by-step methodology of RIA. It is not a shopping model based on gravitational principles but an allocation of expenditure based on observed and predicted shopping behaviour.

Best practice is also recommended on the qualitative assessment of shopping centres using an appraisal framework based on a scoring system applied to a wide range of indicators and factors of the health of a centre. The average score represents the vitality and viability index for a particular centre. The appraisal framework has been thoroughly tested and has been used successfully in a large number of retail studies. It is a subjective but systematic approach to assessing vitality and viability which allows comparisons between centres and the monitoring of changes in the health of a centre. The interpretation of retail impact must take into account the implications of trade diversion in the light of the qualitative appraisal of town centres. A broad guide to the significance of a predicted level of retail impact can be obtained from the proposed model of significance of impacts.

Government policy guidance also requires other qualitative factors to be used in the assessment of the proposed new retail developments. These factors are:

• qualitative need – taking account of deficiencies in shopping provision, indications of over-trading of existing stores, and consumer demand
• the sequential approach – identifying alternative sites and evaluating sites in terms of their suitability, viability and availability
• accessibility by a choice of means of travel
• impact on travel and car use.

The final judgement about impact on a centre is made by examining the other qualitative factors listed in PPG6 paragraph 4.3, such as risk to the town centre strategy and the effect on future investment in the town centre.

Other lessons to be learnt

The research underlying this book has highlighted the need for better retail statistics which are critical to the information base for retail planning. There is a serious lack of essential information, especially on turnover, since the demise of the Census of Distribution. The local authority survey shows clearly that the availability and quality of information used in RIAs is inadequate. However, local surveys of shopping patterns can help to fill the information gap.

The examination of international experience in the impact of new types of retail development shows that there are few lessons to be learnt for the application of RIA in Britain. There is no clear 'preferred' approach to assessing retail impact in North America or Europe, and the methodology appears to be much less developed than in Britain. Decisions are based mostly on economic impact factors, as in Britain, but there is greater concern about social and environmental impacts. North America has a long history of out-of-town shopping centres, particularly planned shopping centres or malls. There is evidence of serious impact in the form of the decline of town and city centres in the USA, more so than in Canada, but in the USA there is very little control of retail development. The rapid growth of hypermarkets in France has been controlled by legislation, which has recently been strengthened, and there are also strong controls in Germany, but the evidence of impact would suggest that these controls have not been effective enough.

The other main lesson to be learnt, particularly from the views of local authorities, is the need for RIAs to be more independent and objective. There are parallels here between RIA and the guidance on environmental impact assessment (EIA). The government's guide to procedures for EIA states that the aim is for it to be as systematic and objective an account as possible of the effects of a project. The government's good practice guide to EIA also refers to the need for a systematic approach. It says that an environmental statement may be criticised for lack of objectivity if it over-emphasises the positive impacts or suggests that potential adverse impacts can be resolved by the application of mitigating measures. Like an EIA, an RIA should be a professionally prepared and presented document which neither over-emphasises benefits nor understates adverse effects. EIAs are subject to review by independent auditors, and a similar approach should be adopted for RIAs.

In EIA it is an important requirement that the environmental statement must contain a non-technical summary which should provide an accurate and balanced summary of key information contained in the environmental statement. It should describe all the conclusions of the environmental statement and the facts and judgements on which they are based. It should not be treated simply as a public relations document. Developers should be careful to avoid any risk of bias in highlighting the proposal's most favourable features and playing down any adverse effects. RIAs should equally be objective and open to independent scrutiny. Government guidance should make it a requirement for RIAs to be based on guidelines such as those recommended in this book to ensure that these principles are followed.

Current government guidance on RIA is inadequate. PPG6 could have been more explicit about the practical application of RIA, rather than just setting out general principles. The best practice guide in this book is intended to fill this gap. The framework approach recommended is an improvement on current practice which recognises that both quantitative and qualitative assessments are necessary, and that the deficiencies in past approaches to RIA can be overcome by a more practical methodology which meets the policy requirements of PPG6. It has been tested in practice and shown to be a reliable method of predicting the impact of new retail developments. There now appears to be consistency and stability in national policy guidance but further shifts in policy are possible. Even if retail planning policy does change in the future, the framework approach is sufficiently robust to cope with any changes in the guidance. It takes account of the factors which will be relevant to decisions on new retail development in the foreseeable future – regarding need, the impact on the vitality and viability of town centres, and the qualitative issues raised in PPG6. It establishes a consistent approach which is readily understood and can be used for the benefit of local authorities and planning inspectors. This approach is a positive step forward in making informed decisions on proposals for new retail development.

GLOSSARY

bulky goods   non-food items usually sold from retail warehouses, including DIY and hardware goods, electrical goods, carpets and furniture, and other household goods.

catchment area   the area from which trade to a store or centre is attracted; the primary catchment area defines the area from which shoppers tend to use a particular store or centre in preference to other stores or centres.

clawback   the ability of a new retail development to recapture leakage of trade that is lost from a catchment area.

comparison shopping   non-food retail goods or outlets which specialise in the sale of non-food items such as clothing and footwear, bulky goods (see above) and other non-food goods normally sold in town centres.

convenience shopping   food and grocery items or outlets which specialise in the sale of food and groceries, essential household items, news-papers, magazines, tobacco and alcoholic drink.

district centre   groups of shops, separate from the town centre, that usually contain at least one food supermarket or superstore, and non-retail services such as banks, building societies and restaurants.

edge-of-centre   for shopping purposes, a location within easy walking distance (i.e. 200 to 300 metres) of the primary shopping area, often providing parking facilities that serve the centre as well as the store, thus enabling one trip to serve several purposes.

factory outlet centres   groups of shops, usually away from the town centre, specialising in selling seconds and end-of-line goods at discounted prices.

hierarchy of shopping centres   a classification of shopping centres in which the importance of the centre is related to the size of its catchment area.

hypermarket   a large, single-level self-service store selling both food and non-food goods, with a trading floorspace of at least 50,000 square metres.

leakage   the loss of retail spending generated by residents in a catchment area to external centres.

linked trips trips which combine main food shopping with other activities such as non-food shopping and visits to other town centre facilities.

local centre small grouping usually comprising a newsagent, a general grocery store, a sub-post office and occasionally a pharmacy, a hairdresser and other small shops of a local nature.

out-of-centre a location that is clearly separate from a town centre, but not necessarily outside the urban area.

out-of-town an out-of-centre development on a greenfield site or on land not clearly within the current urban boundary.

regional shopping centres out-of-town centres that are generally over 50,000 square metres gross in retail area, typically enclosing a wide range of comparison goods.

residual turnover the turnover remaining in a retail outlet or shopping centre after trade diversion has taken place to a new retail development.

retail warehouses large, single-level stores (usually of at least 10,000 square metres gross floorspace) specialising in the sale of household goods (such as carpets, furniture and electrical goods) and bulky DIY items, catering mainly for car-borne customers and often in out-of-centre locations.

retail parks an agglomeration of at least three retail warehouses.

supermarkets single-level, self-service stores selling mainly food, with a trading floorspace less than 2,500 square metres, usually with car parking.

superstores single-level, self-service stores selling mainly food, or food and non-food goods, usually with more than 2,500 square metres trading floorspace, with supporting car parking.

town centre city, town and traditional suburban centres which provide a broad range of facilities and services and which fulfil a function as a focus for both the community and for public transport. It excludes small parades of shops of purely local importance.

trade diversion the loss of trade from an existing store or centre as a result of a new retail development taking place, usually measured as a percentage of its turnover before the opening of the new development.

trade draw the percentage of the turnover of a new retail development which is drawn from a particular part of its catchment area, usually a travel time isochrone or a geographical sub-area.

yield the current annual income from a property, expressed as a percentage of the property's freehold price.

BIBLIOGRAPHY

Alty, R., Mackie, S., Moseley, J. and Rainford, P. (1979) 'The South Yorkshire shopping model', Unit for Retail Planning Information, report U12, Reading: URPI Ltd.

Arnold, S. (1995) 'Discount growth threatens to undercut the high street', *Planning*, 1134: 22–23.

Arnold, S. (1998) 'Proof that need is not an issue', *Planning*, 20 November: 10.

Baldock, J. (1994) 'Pecking order principles', *Planning Week*, 46: 212–13.

Baldock, J. (1996) 'Planned action needed to break fourth retail wave', *Planning*, 1170: 28–29.

Baldock, J. (1998) 'Factory outlets: a year in town centre management', *Property Week*, p. 37.

Batty, M. (1985) in M.J. Breheny and A.J. Hooper, *Rationality in Planning: Critical Essays on the Role of Rationality in Urban and Regional Planning*, London: Pion.

Batty, M. (1997) editorial, 'The retail revolution', *Environment and Planning B: Planning and Design*, 24: 1.

Batty, M. and Saether, A. (1972) 'A note on the design of shopping models', *Journal of the Royal Town Planning Institute*, 58: 303–306.

BDP Planning and Oxford Institute of Retail Management (1992) 'The effects of major out of town retail development: a literature review for the Department of the Environment', London: HMSO.

Begg-Saffar, M. and Begg, H. (1996) 'Royer reforms tighten grip on competition for small traders', *Planning*, 1180: 21.

Bennison, D.J. and Davies, R.L. (1980) 'The impact of town centre shopping schemes in Britain: their impact on traditional retail environments', *Progress in Planning*, 14: 1–104.

Booton, C. (1994) 'Factory outlet centre faces the inquiry test', *Planning*, 1079: 8–9.

Borchert, J.G. (1988) 'Planning for retail change in The Netherlands', *Built Environment*, 14: 22–37.

Braithwaite, J. (1997) 'Scottish shopping guidance still wanting', *Town and Country Planning*, 66: 22–24.

Breheny, M.J. (1983) 'A practical view of planning theory', *Environment and Planning B*, 10: 101–115.

Breheny, M.J. (1993) 'Planning the sustainable city region', *Town and Country Planning*, 62: 71–75.

Breheny, M.J. and Hooper, A.J. (1985) *Rationality in Planning: Critical Essays on the Role of Rationality in Urban and Regional Planning*, London: Pion.

Breheny, M.J., Green, J. and Roberts, A.J. (1981) 'A practical approach to the assessment of hypermarket impact', *Regional Studies*, 15, 6: 459–474.

Bridges, M.J. (1976) 'The York Asda: a study of changing patterns around a superstore', Centre for Urban and Regional Research, University of Manchester.

Bromley, R.D.F. and Thomas, C.J. (1993) *Retail Change: Contemporary Issue*, London: UCL Press.

Burke, T. and Shackleton, J.R. (1996) *Trouble in Store? UK Retailing in the 1990s*, London: Institute of Economic Affairs.

Burt, S.L. (1985) 'The Loi Royer and hypermarket development in France: a study of public policy towards retailing', PhD thesis, University of Stirling.

Carey, R.J. (1988) 'American downtowns: past and present attempts at revitalisation', *Built Environment*, 14: 47–59.

Carlson, H.J. (1991) 'The role of the shopping centre in US retailing', *International Journal of Retail and Distribution Management*, 19, 6: 13–20.

CB Hillier Parker and Savell Bird Axon (1998) 'The impact of large foodstores on market towns and district centres', DETR research report, London: The Stationery Office.

Chase, M. and Drummond, P. (1993) 'Shopping after the millennium', Town and Country Planning Summer School, University of Lancaster: 16–19.

Cordey-Hayes, M. (1968) 'Retail location models', CES working paper 16, Centre for Environmental Studies.

The Data Consultancy (1998) 'Consumer retail expenditure estimates for small areas: explanatory volume, 1995 expenditure'.

The Data Consultancy (1999a) 'GB business-based retail expenditure estimates and price indices', Information brief 99/1.

The Data Consultancy (1999b) 'UK goods-based retail expenditure estimates and price indices', Information brief 99/2.

The Data Consultancy (1999c) 'Markets', special report by The Data Consultancy.

Davey, J. (1999a) 'Bluewater opens floodgates for rateable value revaluation', *Property Week*, 23 April: 13–14.

Davey, J. (1999b) 'City decentered', *Property Week*, 5 November: 12–13.

Davies, E. (1995) 'Understanding government retail policy in a hostile environment', report, 3: 29–30.

Davies, K. and Sparks, L. (1989) 'The development of superstore retailing in GB 1960–1986: results from a new database', transactions of the Institute of British Geographers, 14: 74–89.

Davies, L. (1996) 'Different sections to the same town', *Planning Week*, 4: 11.

Davies, R.L. (1976) *Marketing Geography with Special Reference to Retailing*, London: Methuen.

Davies, R.L. (1984) *Retail and Commercial Planning*, London: Croom Helm.

Davies, R.L. (1986) 'Retail planning in disarray', *The Planner*, 72, 7: 20–22.

Davies, R.L. and Howard, E. (1988) 'Issues in retail planning within the UK', *Built Environment*, 14: 7–21.

Davies, R.L. and Kirby, D.A. (1980) 'Retail organisation', in J.A. Dawson (ed.) *Retail Geography*, London: Croom Helm.

Davies, R.L. and Rogers, D.S. (eds) (1984) *Store Location and Store Assessment Research*, Chichester: J. Wiley and Sons.

Dawson, J.A. (1980) *Retail Geography*, London: Croom Helm.

Dawson J.A. (1981) 'Shopping centres in France', *Geography*, 66, 2: 143–146.

Dawson J.A. (1983) *Shopping Centre Development*, Harlow: Longman.

Dawson, J.A. and Lord, J.D. (eds) (1985) *Shopping Centre Development: Policies and Prospects*, London: Croom Helm.

Dawson, J.A. and Sparks, L. (1985) 'Information sources and retail planning', University of Stirling, Department of Business Studies, Working paper 8505.

Dawson, J.A. and Sparks, L. (1986) 'Information provision for retail planning', *The Planner*, 72, 7: 23–26.

Dawson, J.A., Granby, D.M. and Schiller, R. (1988) 'The changing high street', *Geographical Journal*, 154: 1–22.

de Vaus, D.A. (1996) *Surveys in Social Research*, London: UCL Press.

Denison, N. (1996) 'Note locks the door on rival retail plans', *Planning*, 1178: 8–9.

DoE (1988) PPG6, 'Major retail development', London: HMSO.

DoE (1992a) PPG1, 'General policy and principles', London: HMSO.

DoE (1992b) Circular 19/92, 'Development plans and consultation direction', London: HMSO.

DoE (1993a) PPG6, 'Town centres and retail development', London: HMSO.

DoE (1993b) Circular 15/93, 'Shopping development direction', London: HMSO.

DoE (1994) PPG13, 'Transport', London: HMSO.

DoE (1995a) ' "Shopping centres and their future", the government's response to the House of Commons Select Committee on the Environment', Cm 276, London: HMSO.

DoE (1995b) Revised PPG6, 'Town centres and retail development', consultation draft, London: HMSO.

DoE (1996) Revised PPG6, 'Town centres and retail development', London: HMSO.

DoE (1997) Revised PPG1, 'General policy and principles', London: HMSO.

DETR (1995) 'Preparation of environmental statements for planning projects that require environmental assessment: a good practice guide', London: HMSO.

DETR (1998) British Council of Shopping Centre annual conference, 5 November 1998, speech by Richard Caborn, Minister for the Regions, Regeneration and Planning, press release.

DETR (1999a) 'Planning applications for shopping and leisure schemes should be assessed on the basis of need', press release.

DETR (1999b) 'A better quality of life', London: The Stationery Office.

DETR (1999c) Draft revision of PPG13, 'Transport', London: The Stationery Office.

DETR (1999d) 'Town and Country Planning (Development Plans and Consultation) Direction', Circular 7/99, London: The Stationery Office.

DETR (1999e) 'Positive planning: the key to new retail development' – Richard Caborn, press release.

Distributive Trades EDC (1970) 'Urban models for shopping studies', London: NEDO.

Distributive Trades EDC (1971) 'The future pattern of shopping', London: NEDO.

Distributive Trades EDC (1988) 'The future of the high street', London: NEDO.

Doidge, R. (1999) 'Factory outlets', *Property Week* supplement.

Donaldsons (1979) *Caerphilly Hypermarket Study: Year Five*, London: Donaldson and Sons.

Drivers Jonas (1992) *Retail Impact Assessment Methodologies: Research Study for the Scottish Office*, Edinburgh: the Scottish Office.

Drysdale, R. (1995) 'Scottish retail guidelines sets alarm bells ringing', *Planning*, 1123: 26–27.

Duffill, C. (1995) 'Adverse impact figure could face a bumpy ride', *Planning*, 1115: 8–9.

Eade, C. (1999) 'Wonder Wal', *Property Week*, 5 November: 34–36.

Eastman, C. (1995) 'PPG13 and retailing: travel to Safeway superstores', 7th annual TRICS conference.

England, J.R. (1996) 'Planners take dim view of retail impact studies', *Planning*, 1171: 24–25.

England, J.R. (1997) 'Retail impact assessment: a critical examination of its application in the planning process', PhD thesis, University of Newcastle-upon-Tyne.

English Historic Towns Forum (1997) 'Retail guidance', Report no. 40.

Faludi, A. (1973) *A Reader in Planning Theory*, Oxford: Pergamon.

Fernie, J. (1995) 'The coming of the Fourth Wave: new forms of retail out-of-town development', *International Journal of Retail and Distribution Management*, 23, 1: 4–11.

Fernie, S. (1996) 'The future for factory outlet in the UK: the impact of changes in planning policy guidance on the growth of a new retail format', *International Journal of Retail and Distribution Management*, 24, 6: 11–21.

Foley, D.L. (1973) 'British town planning: one ideology or three?', in A. Faludi, *A Reader in Planning Theory*, Oxford: Pergamon.

Friedmann, J. (1987) *Planning in the Public Domain*, Princeton, New Jersey: Princeton University Press.

Gayler, H.J. (1984) *Retail Innovation in Britain: The Problems of Out-of-town Shopping Centre Development*, Norwich: Geo-Books.

Gayler, H.J. (1989) 'The retail revolution in Britain', *Town and Country Planning*, 58: 277–280.

Ghosh, A. and McLafferty, S. (1991) 'The shopping center: a restructuring of post-war retailing', *Journal of Retailing*, 67, 3: 253–267.

Gibbs, A. (1981) 'An analysis of retail warehouse planning inquiries', URPI Report U22, Unit for Retail Planning Information.

Gibbs, A. (1986) 'Retail warehouse planning inquiries', URPI Report U28, Unit for Retail Planning Information.

Gibbs, A. (1987) 'Retail innovation and planning', *Progress in Planning*, 27, 1: 1–67.

Goddard, C. (1999) 'A policy for taking heed of need', *Planning*, 19 February: 13.

Goss, J. (1993) 'The magic of the mall: an analysis of form, function and meaning in the contemporary retail built environment', *Annals of the Association of American Geographers*, 83, 1: 18–47.

Greater London Council (1980) 'The impact of Brent Cross', Reviews and Studies Series no. 2.

Grimley, J.R.E. (September 1993) 'The national foodstore department report'.

Guy, C.M. (1977) 'A method of examining and evaluating the impact of major retail developments upon existing shops and their users', *Environment and Planning A*, 9: 491–504.

Guy, C.M. (1984) 'The estimation of retail turnover for planning purposes', *The Planner*, 70, 5: 12–14.

Guy, C.M. (1987) 'The assessment of retail impact', *The Planner*, 73, 12: 31–34.

Guy, C.M. (1988) 'Retail planning policy and large grocery store development: a case study in South Wales', *Land Development Studies*, 5: 31–45.

Guy, C.M. (1991) 'Spatial interaction modelling in retail planning practice: the need for robust statistical methods', *Environment and Planning B*, 18: 191–203.

Guy, C.M. (1994a) 'Whatever happened to regional shopping centres?', *Geography*, 79, 4: 293–312.

Guy, C.M. (1994b) *The Retail Development Process: Location, Property and Planning*, London: Routledge.

Guy, C.M. (1994c) 'Grocery store saturation – has it arrived yet?', *International Journal of Retail and Distribution Management*, 22, 1: 3–11.

Guy, C.M. (1996) 'Grocery store saturation in the UK – the continuing debate', *International Journal of Retail and Distribution Management*, 24, 6: 3–10.

Hague, C. (1991) 'A review of planning theory in Britain', *Town Planning Review*, 62, 3: 295–310.

Hall, P. (1988) *Retail Geography*, London: Croom Helm.

Hall, P. and Breheny, M. (1987) 'Urban decentralisation and retail development: Anglo-American comparison', *Built Environment*, 13, 4: 244–261.

Hallsworth, A. (1990a) 'More home thoughts from abroad', *Town and Country Planning*, 59: 51–53.

Hallsworth, A. (1990b) 'The lure of the USA: some further reflections', *Environment and Planning A*, 22: 551–558.

Halman, G. (1998) 'It's not easy to find the edge of centre', *Planning*, 30 October 1998: 14–15.

Hansard (1999) 24 June 1999, Column 1351.

Hayton, K. (1996) 'Centre blight or boost in shopping guidelines?', *Planning*, 1172: 8–9.

Healey, P. (1983) 'Rational method as a mode of policy formulation and implementation in land use policy', *Environment and Planning B*, 10: 19–39.

Healey, P. (1990) 'Understanding land and property development processes: some key issues', in P. Healey, and R. Nabarro (eds) *Land and Property Development in a Changing Context*, Aldershot: Gower.

Healey, P. (1991a) 'Debates in planning thought', in H. Thomas, and P. Healey (eds) *Dilemma of Planning Practice*, Aldershot: Avebury.

Healey, P. (1991b) 'Models of the development process: a review', *Journal of Property Research*, 8: 219–238.

Healey, P. (1992a) 'Planning through debate', *Town Planning Review*, 63, 2: 143–162.

Healey, P. (1992b) 'The reorganisation of state and market in planning', *Urban Studies*, 29, 3–4: 411–434.

Healey, P. (1992c) 'An institutional model of the development process', *Journal of Property Research*, 9: 33–44.

Healey, P. and Shaw, T. (1993) 'Planners, plans and sustainable development', *Regional Studies*, 27, 08/01: 769–776.

Healey and Baker (1995) 'Measuring vitality and viability: a critical analysis of the tests of PPG6', a report for Tesco Stores Ltd.

Hogarth-Scott, S. and Rice, S.P. (1994) 'The new food discounters – are they a threat to the major multiples?', *International Journal of Retail and Distribution Management*, 22, 1: 20–28.

Holt, G. (ed.) (1998) *Development Control Practice*, Gloucester: Ambit Publications.

House of Commons, Fourth report (1994) 'Shopping centres and their future', London: HMSO.

House of Commons, Fourth report (1997) 'Shopping centres', London: HMSO.

Howard, E.B. (October 1986) 'Measuring the impact of the big schemes', *Town and Country Planning*, 55: 282–284.

Howard, E.B. (1988) 'Impact studies: how valuable are they?', Oxford Institute of Retail Management, Research paper A17.

Howard, E.B. (1989) *Prospects for Out of Town Retailing: The Metro Centre Experience*, Oxford Institute of Retail Management.

Howard, E.B. and Davies, R.L. (1992a) 'Meadowhall: the impact of one year's trading', Oxford Institute of Retail Management, Research paper D9.

Howard, E.B. and Davies, R.L. (1992b) 'The impact of the Metro Centre', Oxford Institute of Retail Management, Research paper A27.

Howard, E.B. and Davies, R.L. (1993) 'The impact of regional out-of-town centres: the case of the Metro Centre', *Progress in Planning*, 40, 2: 89–151.

Inman, J. (1995) 'Assessing retail impact', *The Scottish Planner*.

IPD (1996) 'IPD town centre league table', IPD special report.

Jensen-Butler, C. (1972) 'Gravity models as planning tools: a review of theoretical and operational problems', *Geografiska Annaler B*, 54: 68–78.

Johnston, B. (1987) 'Megastore madness', *Housing and Planning Review*, 42, 3: 4–26.

Jones, P. (1989) 'Regional shopping centres', *Town and Country Planning*, 58: 280–283.

Jones, P. (1995a) 'Factory outlet shopping developments', *Geography*, 80, 348: 277–280.

Jones, P. (1995b) 'Factory outlet shopping centres and planning issues', *International Journal of Retail and Distribution Management*, 23, 1: 12–17.

Jones, P. and Vignali, C. (1993) 'Factory outlet shopping centres', *Town and Country Planning*, 62: 240–241.

Kirby, D.A. (1986) *Retailing and Retail Planning: A Guide to Sources and Information*, Stamford: Capital Planning Information.

Kivell, P.T. and Shaw, G. (1980) 'The study of retail location', in J.A. Dawson (ed.) *Retail Geography*, London: Croom Helm.

Klosterman, R.E. (1985) 'Arguments for and against planning', *Town Planning Review*, 56: 5–20.

Kulke, E. (1996) 'American planning in the 1990s: evolution, debate and challenge', *Urban Studies*, 33, 4–5: 649–671.

Langston, P., Clarke, G.P. and Clarke, D.B. (1997) 'Retail saturation, retail location and retail competition: an analysis of British grocery retailing', *Environment and Planning A*, 29: 77–104.

Lavery, C. (1987) 'Retailing: past, present and future', *Housing and Planning Review*, 3, 42: 4–26.

Lee Donaldson Associates (1991) 'Superstores appeals', Review three: Research study five.

Lucas, R. (1995) 'Retail planning – past, present and future', *Report*, 3: 16–18.

McCallum, D. (1995) 'Retail impact studies – a beginner's guide', *Report*, 3: 24–25.

McLoughlin, J.B. (1969) *Urban and Regional Planning: A Systems Approach*, London: Faber and Faber.

Mills, E. (1974) 'Recent developments in retailing and urban planning', PRAG technical paper, Planning Research Applications Group.

Moser, C.A. and Kalton, G. (1993) *Survey Methods in Social Investigation*, Aldershot: Dartmouth Publishing.

Moss, N. and Fellows, M. (1995) 'The future for town centres', *Estates Gazette*, 9506: 141–143.

National Retail Planning Forum (1999) 'A bibliography of retail planning', Institute of Retail Studies, University of Stirling.

Noel, C.L. (1989) 'Retail impact assessments: a practical appraisal', Volume 1, Working paper no. 118, Oxford: Oxford Polytechnic.

Noel, C.L. (1990) 'Retail impact assessments: a practical appraisal', Volume 2, Working paper no. 119, Oxford: Oxford Polytechnic.

Norris, S. (1992) 'The "return" of impact assessment: the rise of the regional shopping centre and the return of impact assessment methods in the UK', PhD thesis, University of Reading.

Norris, S. and Jones, P. (1993) 'Retail impact assessment – the future?', *Estates Gazette*, 9304: 84–87, 88.

Openshaw, S. (1973) 'Insoluble problems in shopping model calibration when the trip pattern is not known', *Regional Studies*, 7: 367–371.

Pacione, M. (1979) 'The in-town hypermarket: an innovation in the geography of retailing', *Regional Studies*, 13, 1: 15–24.

Pal, J. (1996) 'Discount stores faced with double standards', *Planning*, 1178: 9.

Parker, A. (1995) 'Market towns and food stores: a new policy approach', *Estates Gazette*, 9503: 108–109.

Parsons, D. and Sherman, P. (1994a) 'Counter revolution', *Planning Week*, 2, 9: 18–19.

Parsons, D. and Sherman, P. (1994b) 'Shopping around', *Planning Week*, 2, 17: 14–15.

Plowden, S. and Hillman, M. (1995) 'Shopping centres – whither or wither?', *Town and Country Planning*, 64, 1: 3–4.

Pope, M. (1996) 'Locational trends: a trader's view', Planning and Environmental Training (PET) conference, 'Shopping', Doncaster.

Potter, R.B. (1982) *The Urban Retailing System, Cognition and Behaviour*, Aldershot: Gower.

Property Week (1998) 'Border controls', 30 October: 32–33.

Property Week (1999a) 'The retail guide to Europe and the Middle East' (report in association with Healey and Baker).

Property Week (1999b) 'Freeport lifts off on the continent', 19 November: 13.

Raggett, B. (1994a) 'Acting in the interests of the public realm', *Planning Week*, 18, 2: 10–11.

Raggett, B. (1994b) 'The revised PPG6 – can you spot the difference?', Planning and environmental training seminar, 'New trends in shopping after PPG6', London.

Raggett, B. (1996) 'The realities of retailing – aims, background themes and signals from Marsham Street', PET conference, 'The realities of retailing', London.

Rees, J. (1987) 'Perspectives on retail planning issues', *Planning Research and Practice*, 2: 3–8.

Reeves, S. (1996) 'Place for outlet centres in the retail hierarchy?', *Planning*, 1174: 10.

Reynolds, J. and Howard, E. (1992) 'The UK regional shopping centre: the challenge for public policy', OXIRM Research paper A28.

Rhodes, T. and Whitaker, R. (1967) 'Forecasting shopping demand', *Journal of the Town Planning Institute*, 53: 188–192.

Roberts, A. (1982) 'Superstore and hypermarket impact analysis', *The Planner*, 1, 68: 8–11.

Robson, D. (1987) 'Trading impact of a superstore – a London case study and its implications', *The Planner*, 10, 73: 16–18.

Roebuck, S. and Goddard, C. (1993) 'Choice and opportunities for all in new shopping guidance?', *Planning*, 1028: 16–17.

Roger Tym & Partners (1986) 'Greater Manchester shopping study', consultants' report.

Roger Tym & Partners (1993) *Merry Hill Impact Study*, London: HMSO.

Roger Tym & Partners (1997) 'Somerfield linked shopping trips survey' (research report for Somerfield Stores Ltd).

Rogers, David (1991) 'An overview of American retail trends', *International Journal of Retail and Distribution Management*, 19, 6: 3–12.

RTPI (1988) 'Planning for shopping into the 21st century', RTPI Retail Planning Working Party.

RTPI (1994) 'Shopping centres and their future', submission of evidence to the House of Commons Select Committee on the Environment on its enquiry, RTPI.

Schiller, R. (1981) 'Superstore impact', *The Planner*, 2, 67: 38–40.

Schiller, R. (1986) 'Retail decentralisation: the coming of the third wave', *The Planner*, 72, 7: 13–15.

Schiller, R. and Reynolds, J. (1991) 'Ranking British centres', *Estates Gazette*, 9117: 60–61, 81.

Scott, A.J. and Roweis, S.T. (1977) 'Urban planning in theory and practice: a reappraisal', *Environment and Planning*, 9: 107–1119.

Scottish Development Department (1986) National planning guidelines 1986, 'Location of major retail developments', the Scottish Office.

Scottish Development Department (1996) NPPG 8, 'Retailing', the Scottish Office.

Scottish Grocers Federation (1994) 'Discount retailer survey'.

Self, P. (1993) *Government by the Market: The Politics of Public Choice*, Basingstoke: Macmillan.

Selman, P. (1995) 'Local sustainability', *Town Planning Review*, 66, 3: 287–302.

Shaw, G. (1987) 'Institutional forces and retail change: a case study of metropolitan Toronto', *Geoforum*, 18: 361–369.

Shepherd, I.D. and Thomas, C.J. (1980) 'Urban consumer behaviour', in J.A. Dawson (ed.) *Retail Geography*, pp. 18–94, London: Croom Helm.

Sherman, P. and Dossett, M. (1995) 'Shopping around', *Planning Week*, 4 May, 3, 18: 11–13.

Simmie, J. and Sutcliffe, A. (1994) 'The death and life of town centres', *Planning*, 1081: 6–7.

Smith, G. (1994) 'Vitality and viability of town centres', *Journal of Planning and Environmental Law*, pp. 91–106.

Sorensen, A.D. (1983) 'Towards a market theory of planning', *The Planner*, 69, 3: 78–80.

Sorensen, A.D. and Day, R.A. (1981) 'Libertarian planning', *Town Planning Review*, 52, 4: 390–402.

Sparks, L. (1996) 'The census of distribution: 25 years in the dark', *Area*, 28, 1: 89–95.

Stansbury, M. (1994) 'The future of the traditional town centre', RTPI northern branch seminar, 'Trading places', Durham.

Stathers, D. (1996) 'The retailer's view – how to make town centres perform better', PET conference, 'The realities of retailing', London.

Stockdale, J. (1993) 'Whitehall yields to the flow on centre diagnosis', *Planning*, 1034.

Tapley, A. (1993) 'Discount operators ready to ride the next retail wave', *Planning*, 1015.

Teitz, M.B. (1996) 'American planning in the 1990s: evaluation, debate and challenge', *Urban Studies*, 33, 4–5: 649–671.

TEST (1989) 'Trouble in store: retail locational policy in Britain and Germany', report by TEST for the Anglo-German Foundation.

Thomas, K. (1992) 'A case study of the city of Eugene, Lane County, Oregon, USA', Working paper no. 135, Oxford: Oxford Polytechnic School of Planning.

Thomas, C. and Bromley, R. (1995) 'Retail decline and the opportunities for commercial revitalisation of small shopping centres (a case study in South Wales)', *Town Planning Review*, 66, 4: 431–452.

Thorpe, D. (1994) 'The balance between town centre shopping and out of town development (a personal view)', Planning and environmental training seminar, 'New trends in shopping after PPG6', London.

Thorpe, D., Shepherd, P.M. and Bates, P. (1976) 'Food retailers and superstore competition: a study of short term impact in York, Northampton and Cambridge', Retail Outlets Research Unit, Manchester Business School, Research report no. 25.

Tiffin, R. (1996) 'Weighing anchor roles in town centre futures', *Planning*, 1160.

University of Manchester (1964) 'Regional shopping centres: a planning report on north west England', University of Manchester, Department of Town and Country Planning, report of an investigation into a proposal for a regional shopping centre at Haydock, Lancashire.

Urban Management Initiatives (1999) 'The Lockwood survey', Huddersfield.

URBED (1994) 'Vital and viable town centres: meeting the challenge', research report for the Department of the Environment, London: HMSO.

URPI (1986) *MARKETS: A New Retail Planning Tool*, Reading: URPI.

URPI (1995) 'Derivation and use of URPI consumer retail expenditure estimates', Information brief 95/1, URPI.

URPI (1998a) 'UK goods-based retail expenditure estimates and price indices', Information brief 98/2, The Data Consultancy.

URPI (1998b) 'Trends in furniture, electrical and do-it-yourself goods expenditure', Information brief 98/3, The Data Consultancy.

Valuation Office (1998) *Property Market Report*, Autumn, London: the Valuation Office.

Wade, B. (1983) 'Retail planning without data', *The Planner*, 69, 1: 26–28.

Westlake, T. and Smith, M. (1994) 'Facing the fourth wave of retail development?', *Town and Country Planning*, 12, 63: 334–335.

Westlake, T. and Forsburg, H. (1996) 'Sweden opens up to out of town', *Town and Country Planning*, 65: 27–28.

Williams, H. (1995) 'Food shopping – where next?', *Report*, 3: 27–28.

Williams, J.J. (1991) 'Meadowhall: its impact on Sheffield city centre and Rotherham', *International Journal of Retail and Distribution Management*, 19, 1: 29–37.

Williams, R.H. and Wood, B. (1994) *Urban Land and Property Markets in the United Kingdom*, London: UCL Press.

Willis, K.G. (1980) *The Economics of Town and Country Planning*, London: Granada Publishing.

Worrall, J. (1994) 'PPG6 in action', *Planning Week*, 2, 25: 16–17.

Wrigley, N. (1991) 'Commentary', *Environment and Planning A*, 23: 1537–1544.

Wrigley, N. (1992) 'Commentary', *Environment and Planning A*, 24: 1521–1527.

Zentes, J. and Schwartz-Zanetti, W. (1988) 'Planning for retail change in West Germany', *Built Environment*, 14: 38–46.

INDEX

Milton Keynes UK
Ingram Content Group UK Ltd.
UKHW031532071024
449327UK00005B/117